May I Ask a Technical Question?

Volume I: Questions about digital reliability each of us should ask.

Jeff Krinock and Matt Hoff

ISBN-10: 0998331309

ISBN-13: 9780998331300

CONTENTS

ACKNOWLEDGMENTS

We wish to thank the many folks who helped make this book possible. The research for this book took many years to complete and involved several turns of direction. Early in that research, my late friend, Thomas Beltzer (*Parcheesi Blues*, Livingston Press, 2007), provided some chilling anecdotes about censorship in this country that gave us an initial and vitally important vector. We also had countless informative conversations with folks of all walks of life—for who among us has not been touched by digital technology?

We wish to thank Jack Herbein, Jeff's occasional weightlifting partner, for his willingness to discuss some of the factors and thinking that go into complex double- and triple-redundant systems. While Jack's experience at Three Mile Island in 1979 was not particularly reliant on digital technology, we know that redundant safety systems are integral to the reliability of many digitally controlled systems. As massive airliners and increasingly tech-laden cars and trucks gain more autonomy, we value the lessons learned from people like Jack, who witnessed failure firsthand in situations when failure was not an option.

Many friends and family helped us review the chapters that follow: Mark Hoff, John Fortunato, Michael Dell, Dr. James Demaio (former USAF flight surgeon), Christie Miller, Ron and Kathy Krinock, and Jeff's wife Pam Krinock. Thank you all for helping us review, and in some cases, decide to dispose of early chapters and drafts. With a variety of backgrounds, from military, federal law enforcement, fiction writing, editing, medicine, and banking, your input remains invaluable, and we thank you all for your patience and support.

Where we've fallen short or have left confusion on these pages, our reviewers are not to blame! In the last few years, we've witnessed some of the fastest-moving change the digital world has ever seen. Some of this is due to whistleblowers and their kin (William Binney, Thomas Drake, and Edward Snowden) who have unveiled an entire world of digital activity that touches every single human being. The pervasive and cloaked essence of digital technology can be seen in the fact that this digital surveillance world was built and established almost completely unnoticed. In any case, trying to keep our in-progress drafts current with unfolding world events presented a significant challenge. The authors alone are to blame in any instances in which we failed to meet that challenge.

Finally, we'd like to thank an unusual and largely invisible group—the thousands of secretaries, admin assistants, clerks, and everyday workers who struggle to understand how they can both be responsive to the needs of those they serve (i.e., employers, clients, customers, patients, etc.) while also being held captive to the caprices of computers and digital technology. To say this "double bind" of our era is ubiquitous would be a gross understatement. In many ways, their daily struggles, evidenced in countless handwritten notes with messages like "Computer is down...thanks for your patience," provided the impetus for beginning this book.

PREFACE

A Possible Variant of the Gospel, Had it Been Tweeted:
"Jesus wept. Judas went out and hung himself...go thou and do likewise."

Some years ago, we had a paper published as a journal feature article, and while that was an honor, our joy quickly turned to embarrassment upon reading the journal editor's introduction. Unfortunately, the editor completely missed the point we had tried to make and practically reversed our intended message. We did not go out and hang ourselves in accordance with the "tweeted Bible"—we've read the non-tweeted version of the Gospel and realize there are some passages between Judas's self-destruction and the command to "do likewise" that deserve consideration. But the public embarrassment of having our writing completely misunderstood reminded us that the speed with which we're now expected to read, digest, respond, comment, and re-tweet is, in some ways, antithetical to the very medium of writing itself. We could not fault the journal editor for misunderstanding our intended thesis; he lives and works with the same time pressures and unreasonable deadlines most of us endure today.

In putting this book together, we struggled in determining its intended audience. Figuring out who will read your work versus who you *would like* to read your work and, even more quixotically, trying to know in advance how you can make a positive impact in people's lives via your printed word is a tall order. Somewhere in there you hopefully find a working compromise between those various groups and your goals, but arriving at that compromise is trickier than some might think. The printed word, unlike so many newer types of media, is meant to last in a (mostly) stable form. And if it lasts long enough, future audiences—even near-future audiences—may change in ways authors cannot fully project.

However, a clear-eyed, forward-looking author like Alexis de Tocqueville, whom we quote extensively in the pages that follow, certainly would not have been surprised to know that many of his future American readers (i.e., today's learners) would read his works in isolation, without the presumed benefit of public and/or academic discussion; this despite the fact he lived and wrote during an era that made great use of public meetings, discussions, and forums—of the face-to-face variety.

De Tocqueville foresaw social isolation and the decay of future opportunities to enjoy deep academic experiences in face-to-face meetings. And he anticipated this change coming to America decades before the invention of the telephone, not to mention radio, television, the Internet, and the other technologies that allow us to receive an author's message without the benefit of the discussion and the reflection that often accompany social and academic interactions in face-to-face settings.

We can't claim de Tocqueville-level insight into how audiences for this book might change. At the dawn of the Internet, which was not so long ago at all, people uploaded complete books with zeal, making classic works available to people around the world. While that trend has not necessarily stopped, it's difficult to miss the changes that have come with our adoption of "social media." Countless young folks currently read and communicate in variants of the written word that all too often lead to inevitable misunderstandings akin to the tragically humorous title of this preface. When in the history of human language have we so willingly truncated thoughts, concepts, phrases, and, most importantly, our opportunities to understand one another at deeper, more complete levels? A wayward "text message" today can turn an erstwhile BFF into a distant memory as fast as you can hit the *Send* button.

We can debate the role of the printed word or even debate where and how we should read. And we can easily guess that many, if not most, of today's readers would say there is no "should" involved in that debate. Give most readers today a slick Apple or Google device and access to social media, and life is grand. Nonetheless, as a writing partner of mine once noted, you don't teach a child to read, lock him in a library, and expect him to emerge with a PhD in a few years. We believe that similar principles apply to social media. Does anyone really believe unfettered access to tweets and posts will somehow lead to highly elevated thinking and understanding of the world in which we live?

There's little debate that how we employ the written word has changed in recent years. If nothing else, there is certainly an increased element of distance between authors and their readers, and writers find absolutely no guarantees that written works will be read carefully, or that readers will have the opportunity to engage instructors when interpreting the real meaning of the written word. (Ironic—social media simultaneously shortens the time required to communicate, while inadvertently maintaining or extending the distance between the persons communicating. We'll likely explore this phenomenon in a future work).

Michael G. Moore, currently of Penn State University, provided a wonderfully timed thesis about transactional distance in 1973, which was born of his work in studying distance learning; at that time, correspondence courses were a well-established feature in the educational world and one of the relatively few options for instruction that did not involve a classroom setting. Most of the world had functional postal systems, and I recall organizations as technically advanced as the US Air Force using correspondence courses to teach even military concepts; at one point in my Air Force career, I received large, well-written courseware via US mail, completed the courseware at home, and received much-needed credit for continuing education without having access to a live instructor in any way, shape, or form.

Moore, for his part, recognized that instruction, such as correspondence courses, carried out without the benefit of face-to-face interactions with authors or instructors differed from classroom-based instruction. He wanted to understand how various forms of instruction differed, and so he came up with the idea of transactional distance. In 1980, he defined transactional distance as follows: "A psychological and communication space to be crossed; a space of potential misunderstanding between the inputs of instructor and those of a learner."

The first edition of this book will be published digitally. Whether you read this "online" or in a printed format, we hope we are able to bridge the communication space. Our hope is to have that contemplative "conversation" that is part of deep reading, contemplation that goes beyond the giddiness of tweets and LOL text messages. If we succeed in any way, please let us know, or better yet, join the conversation with more books and articles in our chosen genre—cyber-skepticism.

1 INTRODUCTION

Are We There Yet?

Ask Google, and we get a resounding "no" – we haven't scratched the surface of what's to come in the digital world. We have no holographic assistants hovering nearby to answer all our questions and to entertain our kids. We're not wearing disease-detecting contact lenses that administer medicine with uncanny omniscience. Our packages aren't delivered by drones, nor are thousands of driverless cars zooming through the streets.

If we could ask George Orwell or Aldous Huxley if we're there yet, however, we'd likely encounter a very different point of view.

For a less giddy perspective about the role of digital technology in our current culture, ask the woman who was fired from her job because she tweeted something controversial.[1] Ask any pilot about modern cockpits and the risks associated with over-reliance on autopilots. Or look at the research of MIT's Sherry Turkle that shows many of us have started to devote extensive time to managing our online lives at the expense of our real relationships.

On another note, every day we hear about "stolen identities" and gigantic hacking incidents like those at Anthem, one of the biggest health insurance providers in the United States, in which over 80 million people had private information exposed. We have also recently witnessed a very public and embarrassing hack at Sony Pictures, exposing juicy internal emails of high-level executives.[2]

We *have* changed the world with our digital technology. And we'd like to pause the collective back-slapping and high-fiving to recognize that fact and slow down the "Are we there yet?" questions long enough to figure out "Where, exactly, are we headed?" And it might be nice to also get a feel for how much the journey will cost us along the way.

The perpetual nature of the question "Are we there yet?" is rooted in the fact that some believe we will never "be there." Like a billionaire who never turns down a dollar (no offense, billionaires, we hope you're enjoying this book), visionaries constantly tout the latest and greatest by introducing, marketing, and expanding our digital footprints, seemingly without end. If there is a part of life that's possible to digitize, then rest assured those like Kurzweil or Moravec likely endorse the idea. On the flip side, some of the digi-skeptics among us believe we are already there, and exactly where we are may not be the digital utopia we've been promised—at least not if we take note of digital technology's countless unintended consequences all around us.

In his book *To Save Everything, Click Here!* (2013), Morozov makes the case that we need to stop treating "The Internet" as a single salvific entity. He urges us to challenge the solutionist worldview—the view that everything can be improved if we simply digitize it and move it to the Web.

Quite simply, we need to be more careful with how we introduce new technologies. We're moving at a pace with digital technology that does not seem legally or socially sustainable, and we even show in Part II of this book that our physical safety is too often in danger due to imprudent use of digital technology. We are ill-equipped to handle the unintended consequences (or even to manage many of the *intended* consequences) of some of the technologies we're using, as we'll discuss in the chapters that follow.

Why Write?

We have the simplest of reasons for choosing to write a book that encourages people to question digital technology: not enough people *are* questioning technology, and we believe the result is that human freedom and fulfillment are too often placed at risk.

Of course, stating that human freedom and fulfillment are at risk is itself a most subjective assertion. Even our underlying assumption that human freedom is a "good thing" is somewhat subjective. Indeed, in "The Grand Inquisitor," a chapter in arguably the greatest novel ever written, Dostoevsky questions chillingly how much freedom human beings really want.

More recently, the Wachowski brothers questioned the human desire for freedom and fulfillment in their "Matrix" films. Their famous "blue pill or red pill" scene wouldn't work if the film's audience didn't implicitly understand that when given the opportunity to exit a sedated and imprisoning virtual world, many individuals would choose ongoing sedation. Stated another way, the movie's "blue pill" recognizes the fact that many individuals would choose to remain within the shadows of Plato's cave rather than facing painful truths.

Regarding human fulfillment in the "Matrix" films, Agent Smith, speaking for the ruling machines, explains that the machines couldn't even provide their captive humans with the comfort of a "utopian" virtual environment, lest people self-destruct within it.

That said, for us, human freedom and fulfillment remains an unequivocal "good." We choose the red pill.

About that Red Pill...

For never was there anything more unbearable to the human race than personal freedom!

— Spoken by the character Ivan in "The Grand Inquisitor," from *The Brothers Karamazov* by Fyodor Dostoevsky

Our book is not an examination of human freedom, per se; that subject is vast and ancient and concerns far more complex and profound issues than does digital reliability. What we do want to say, however, is that digital reliability *touches human freedom*, and we can see this particularly clearly once we provide ourselves a complete and appropriate definition of digital reliability (which we undertake more fully in the chapters that follow).

For now, we'll begin by saying that digital reliability, as both a concept and a real-world attribute of our tools and systems, should mean far more to us than determining whether an electronic widget fires up when we turn on the switch.

This book is part of the nascent genre of *cyber-skepticism,* and within that genre, the attribute of digital reliability is (or at least, should be) a proverbial rock star. Langdon Winner, who was writing cyber-relevant works long before the genre of cyber-skepticism had a name *or* a rock star, noted that Mary Shelley was the earliest modern writer to tackle the subject of *technical* reliability,[3] which is the direct forerunner of digital reliability. One could argue that all writings questioning technical reliability—digital or otherwise—are ancestors of cyber-skepticism. This would make Shelley one of the founders of the cyber-skeptic genre we embrace.

Written in the 19th century, Shelley's *Frankenstein; or, The Modern Prometheus* used the powerful metaphor of a technology-enabled man (Victor Frankenstein's monster, more commonly referred to in movies and sci-fi spinoffs simply as *Frankenstein*) to wrestle with the concepts of technology taking on (apparently) autonomous capabilities.

Why did Shelley, and a growing number of writers, feel the need—a need the authors of this book certainly share—to question technology? Isn't technology the bearer of gifts such as mighty sailing ships, bridges, trains, automobiles, jets, and the Internet? On a far more intimate level, isn't technology giving us antibiotics, micro-surgery, and genetic manipulation?

Langdon Winner provides an answer about the need to question technology in his own analysis of Shelley's *Frankenstein*. Winner traces a basic three-step process (create, forget, reap consequences) that unfolded in the creation of the Frankenstein monster, a process that seems to be something of a template for how we, in this culture, handle technology in general. Or perhaps we should say it's a template for how we *mis*handle technology.

Winner notes that if we sometimes end up ignoring our need to think about technology, we certainly don't start out that way. He notes: "At the outset, the development of all technologies reflects the highest attributes of human intelligence, inventiveness, and concern." He then drops the other shoe, noting that we don't stay on that track of employing our "highest attributes." In using technology, he notes that we eventually grant ourselves *license to forget.*[4]

However, this act of forgetting our genuine ownership of technology is much more than a gateway to high times with a laptop. Winner notes that *Frankenstein*, as a metaphor for all technology, comes back to its creator demanding a price for having been created and then neglected. The hapless human creator of the technological monster is "baffled, fearful" and "then looks on in surprise as [his technological creation] returns to him as an autonomous force, with a structure of its own, with demands upon which it insists absolutely. Provided with no plan for its existence, the technological creation enforces a plan upon its creator."

And it is this juncture—the point at which our intentional forgetting comes back to us with *its own apparent intent to avenge*—that impacts human freedom. This same phenomenon drives the current authors to contribute to the genre of cyber-skepticism. In short, in matters digital, as in all of human existence, when we forget (or ignore or trample upon) our responsibility to own our actions and creations, we pay a price. In that regard, ignorance truly is *not* bliss.

Winner summarizes our plight, specifically our plight when we willingly forget, as paralleling that of Shelley's protagonist: "[Victor Frankenstein] never moves beyond the dream of progress, the thirst for power, or the unquestioned belief that the products of science and technology are an unqualified blessing for humankind. Although he is aware of the fact that there is something extraordinary at large in the world, it takes a disaster to convince him that the responsibility is his."

And so we write, and we blog, and we try to begin with a foundation that says that identifying a digital device or technology as being *reliable* means that we consider both the consequences that ensue when the device or technology fails *as well as the sometimes obscure ramifications when it functions as planned.* Sadly, we all too often see failures on the evening news. We ignore those images with great difficulty. But an important part of our thesis here is that *successful* implementation of any given technology too often brings with it consequences that we will never see on the evening news, nor do we ever address those issues adequately.

Is caring deeply about digital technology—and considering carefully what ensues even if we *succeed* in building any given technology—an excess of caution? Is it all that novel for our culture to consider the possibility that a properly functioning technology may very well do more harm than a technology that fails? To wit, throughout much of the latter part of the 20th century, an array of thinkers ranging from sociologists to politicians to theologians asked whether the omnipresent threat of annihilation by well-designed and well-functioning nuclear bombs was impacting the health of our growing children. Indeed, could schoolhouse drills involving climbing under desks at the sound of a bell cause a problem or two in the psychology and souls of growing innocents?

Mobile phones are not nuclear bombs. Yet Evgeny Morozov notes, with some disdain, that in recognizing the depth and breadth of the impact of digital-based communications, we have come to a point at which some of us believe foolishly "that we should be dropping iPhones not bombs" to foster democracy and human freedom.

And there's the rub: we cannot, with sanity, attribute massive life-changing and society-moving ability to a technology while simultaneously conducting ourselves as if that same technology is in all cases a benign and benevolent force on the other. Too often in life the savior is also the judge.

While we don't expect our public schools to introduce iPhone drills with students diving under desks at the first smartphone tweet, we do believe we'll all benefit from increasing our commitment to honesty about the impacts of our digital toys and tools. We would very much like to avoid arriving at the point of soul-withering complacence that Langdon Winner describes so vividly:

> It is at this point that a pervasive ignorance and refusal to know, irresponsibility, and blind faith characterize society's orientation toward the technical. Here it happens that men release powerful changes into the world with cavalier disregard for consequences; that they begin to 'use' apparatus, technique, and organization with no attention to the ways in which these 'tools' unexpectedly rearrange their lives; that they willingly submit the governance of their affairs to the expertise of others.[5]

If you'd like to join us in avoiding the "benevolent care" of a modern-day, tech-enabled Grand Inquisitor or the vengeful rampage of an unrecognized *Frankenstein*, please read on. Join us as we try to figure out how to live out the consequences of having chosen the "red pill".

2 WE DIDN'T START THE FIRE

*That which cannot be instrumentalized without destroying its essence
(human life) must regain its autonomy over technology.*[6]

— Jeffrey Herf: "Technology, Reification, and Romanticism"

*Saying that you don't care about the "right to privacy" because you have nothing to hide
is no different than saying you don't care about "freedom of speech" because you have
nothing to say.*[7]

— Ed Snowden: Interview with Jean-Michel Jarre

"Sir, why do you have a knife in your computer bag?" While the setting for
this incident may sound like an American airport, and the X-ray and metal-
detecting equipment were identical to a typical airport's security gear, the
location for this real-life event was a bit more sensitive. My friend, whose
business briefcase had just been cleared by the same X-ray machine, backed
away from me, putting distance between himself and the uniformed security
guard who had just asked me that question in a loud, clear voice. As I
realized the guard's question was quite serious, I had a question of my own
running through my mind: "How much of a digital trail do we need to be
deemed reliable persons?"

That particular day in this story ended well enough, I suppose. Over some
beers that evening, my backpedaling colleague (and my subordinate for the
consulting project we were working—who, coincidentally enough, just
happened to be from my own small hometown) told me, "Dude, I was
backing away thinking, what's this guy [me] all about?"

This story presents several ironies of our current digital era. One particular
irony that struck me on a personal level was the fact that I had arranged all
the details of this trip in order for our consulting team to conduct several
meetings with high-level personnel who are charged with keeping our
country secure; this fact alone should have earned at least a little trust from
my "friends and colleagues" that day. The thought that my friend's first
reaction to a suspect X-ray of my computer bag was to back away
soundlessly while doubting my intent (and other necessary aspects of sanity

and stability) gave me a less-than-warm feeling for the day. But, as things go in corporate America, by beer time that evening I was "dude" again to my friend, and the raucous events earlier in the day faded into the surreal.

My apologies in advance for the mini-resume that follows, but before I explain what really happened that day to place me in such needless jeopardy and unwarranted suspicion, I think it's important to consider and understand the question I ask in the first paragraph of this chapter: In truth—what exactly does it take today for the people administering our digital-encrusted world to grant us a modicum of trust? (I no longer even consider asking for trust as a military veteran and a law-abiding citizen— which in this country is a bridge way too far, apparently now and for the foreseeable future.)

That day, as I led my consulting team to a government agency building in a major eastern city for an afternoon of collaborative meetings, I was sporting a shiny new clearance badge on my chest. Obtaining this badge had been no easy task. When I began my quest for clearance to begin working with this particular US government agency, I was already in possession of an active US Department of Defense security clearance, the same clearance I'd maintained for much of the previous 25 years, beginning with a 12-year stint as an active-duty US Air Force pilot. Later, I again held that DoD clearance as a consultant to NATO, as a project manager for the Army Medical Department at Fort Sam Houston of the US Army, as consultant to US Joint Forces Command in southern Virginia, and, coming full circle from active duty, as a civilian contract consultant and strategic learning planner for the US Air Force.

By the time I sought out this new agency's own unique security clearance— which they explained to me was completely separate from my DoD clearance and was required for me to even enter the buildings of the agency I was newly supporting—I had been a USAF helicopter and fighter pilot, an invited speaker and instructor in support of NATO on three continents and in countless countries, and had been published in a NATO-sponsored book and in multiple US defense-related journals. I had in my possession the previously mentioned DoD clearance and a valid DoD "CAC" card (my second DoD-issued Common Access Card of the previous five years), and had been a speaker at two international security events (alongside two US undersecretaries of defense) and an invited keynote speaker at a convention in one of the newest NATO member nations.

Along the way, I had also collected multiple FAA ratings (including Airline Transport Pilot), driver's licenses in two NATO countries, and an invitation to demonstrate some low-level software I had developed at a meeting of NATO leaders in Belgium. In short, to say that I could not understand why I needed, with that pedigree, yet another government agency clearance just to enter a federal building would be a gross understatement. How many digital databases did the government have showing me as a long-time veteran and a hard-working patriot with deep connections to major organizations that are part of our national defense? And further, having successfully obtained that agency's clearance after attending—in person—an all-day security training session in a federal building in the heart of DC, then to enter, at government invitation, another federal building with that hard-earned clearance badge pinned on my coat jacket and still be subjected to an accusation of knife smuggling in front of my consulting team—this was an outrage difficult to fathom.

As happens so often these days, however, a computer and its software provided an explanation to mollify my outrage. That day as I stood there in front of the X-ray security technician, already tired from a long train ride and rental car extravaganza and knowing a full afternoon of interviews and software research awaited me, no sooner had the words "Sir, why do you have a knife...?" burned their way indelibly into my memory than the blinking red outline of the "knife" in my computer bag faded off the security X-ray technician's computer screen, leaving only the normal picture of tangled wires and laptop outlines that I knew represented my bag's real contents. The technician instantly relaxed. I can't quite say the same for me. He then calmly and professionally explained (sans apology, however) that the computer that assists him in analyzing X-rayed bags and packages entering his building sometimes inserts a faux "threat" into the X-ray picture in order to keep the federal security technicians on their toes.

Heh, heh, heh! The digital joke was on me. Not to worry—the computer system had everything under control; the faux knife disappeared from my bag, and I was allowed into the building to finish the day's consulting work in support of our national security. No matter that for a few anxious moments I'd been accused, to my face and in front of my friends and colleagues, of smuggling a knife into a US federal building inside my computer bag. My friends even got to see the blinking red outline of the "knife" in my bag for a few seconds before the computer decided the training joke was over.

The outline of the phantom knife flashed in red on the computer screen for all the world to see, and I will never forget that outline. If my memory were short, the agency's computer systems would come to the rescue, because, amazingly and very sadly, a few weeks later and hundreds of miles away in a different city, another X-ray security technician for the same agency did the same thing to me in a different federal building. There it was, impossibly, a second time—a blinking red outline of a non-existent knife sitting in my bag amongst my laptop, extra toothbrush, and a package of spearmint chewing gum.

I realize now, in writing about these twin incidents after many a subsequent uneventful trip through airport and federal building security (as "uneventful" as air travel in the US can be these days), that these events impacted me deeply. I was dead tired that day from a long journey when I was suddenly accused of this knife-wielding crime, a crime invented for training purposes by an anonymous computer programmer in some unnamed place. In the midst of my shock, I remember reminding myself to remain completely calm, knowing that, even though I was innocent, an overly enthusiastic response to the omnipotent computer would lead to quite the delightful day of security interviews, pat downs, and who knows what else.

The intentional feigned calm on my part was an act of self-censorship,[8] a concept that plays a huge role in shaping who we are as a people, and a concept that we'll likely see more and more of as we willingly expand the use of digitally enabled control into virtually every aspect of our lives. Peggy Noonan recently wrote an article titled "What We Lose If We Give Up Privacy." In it, she quotes journalist Nat Hentoff:

> [I]f citizens don't have basic privacies—firm protections against the search and seizure of your private communications, for instance—they will be left feeling "threatened." This will make citizens increasingly concerned "about what they say, and they do, and they think." It will have the effect of constricting freedom of expression. Americans will become careful about what they say that can be misunderstood or misinterpreted, and then too careful about what they say that can be understood. The inevitable end of surveillance is self-censorship.[9]

I'd love, for the sake of cauterizing a few wounds, to recount more details of these very real events, including the second technician's much surlier and suspicious response (which may have had something to do with my own quietly seething fury that this impossible event could actually happen a second time). But I'll end this particular story with a few questions: How many times does a computer program falsely accuse someone of a felony in the interest of training? Does this happen daily? Do the technicians explain clearly to the falsely accused just what happened? Do they ever—DHS forbid—*apologize?*

This book, however, is not about security policy, per se. We have questions for those who promote, build, and continue to propagate our digital world—lots of questions—and many of them we formulated during our own experiences with security policies and procedures. The digital reach, however, extends far beyond security policies.

Along the way, we will touch on a bit of philosophy and sociology, and while these subjects are not our main focus, when discussing a technology (i.e., digital technology) that many of our peers contend mimics the very mind of man, we do feel compelled to examine at least some of the underlying reasons for such bold claims.

And yet, one could easily (and somewhat appropriately) ask how anyone could possibly talk about digital technology in the current milieu using anything less than glowing phrases and narratives. For example, I just watched a documentary film about the recent stunning success of digital technology in genomics. The film was replete with interviews of cancer victims given new hope and new life thanks to computers performing mindboggling number crunching in sequencing their genetic material to detect "actionable" flaws in their personal genetic code. Some of the new medications assisting these cancer and disease victims could not possibly be concocted by pharmacists without the sheer computational power of the digital systems the pharmacists and doctors employ.

When the silicon-laced dust settles, we will have recounted some interesting stories about digital technology, including accounts of fantastic successes and equally impressive failures and shortcomings. This book is, if nothing else, a quest for honesty and integrity, beginning with efforts to conceptualize how we view and think about technology and hopefully leading to changes in how we continue to integrate technology, all too often thoughtlessly and carelessly, into every facet of our lives.

We even hope that our approach here will inspire a bit of social-techno metacognition (a neologism, we suspect) that highlights those who have asked us to slow down and consider deeply *what* we are doing with technology and how, where, and *why* we are doing it. Sadly, these individuals are few and far between. Ellul, Marcuse, Winner, Postman, Gatto—relatively few in the world of technology know these names, let alone their insightful works. True, most know the names of Huxley and Orwell—even typically literature-deprived computer programmers and analysts know of these writers—and the contributions of those two eminent dystopians have had a significant influence on *social* metacognition. As we expand beyond

Huxley and Orwell, however, we find works that examine questions we believe all of our contemporaries should ask. In the digitally dependent field of artificial intelligence, for instance, we would ask why we want to mimic the mind and the actions of man via digital machinery—given that humans, notoriously or honorifically, depending upon one's point of view, are not machines.

In this book, much of the work that follows is nothing more sophisticated or complex than real-time examinations of technology-influenced events and incidents. Our hope is to deepen our collective understanding of the "social and historical forces...responsible for technology's 'autonomy' and ... [determine] what, if any sort of social action could regain control of it."[10] Our efforts, of course, may fail. And they may even fail for the same reason Jeffrey Herf dubiously labels Langdon Winner's attempts at "epistemological Luddism" (described as "the dismantling of technology to see how it works and to demystify its operations") as "inadequate."[11] Winner, in his flagship work *Autonomous Technology*, examines our attitudes toward technology and concludes that technology establishes the "central agenda of politics and determines to a great extent the nature of solutions .... To be commanded, technology must first be obeyed." This is a powerful claim, but he uses meticulous and detailed research as well as examples from literature to show that his (and our) collective reification of technology is accurate; that is, he contends, quite well, that we do indeed grant technology a *Frankenstein*-like life and volition of its own.

If there is a continuum of attitudes about the virtue of digital technology—no, "virtue" is too strong and too human an adjective, and we're not ready to inadvertently support the reification of technology through careless wording—better said: If we maintain a continuum of *acceptability* of technology, perhaps its two poles are Luddism on the one hand and what Evgeny Morozov describes as "Internet solutionism" on the other.[12] At both extremes and at all points of that continuum, we find writers who accept the notion of reified and even autonomous technology. Some openly encourage the *furtherance* of research in the direction of autonomous technology.[13] Our own reading of Winner is that while he wants us to recognize how powerful our belief is in the acceptance of autonomous technology (as evidenced by his inclusion of an entire chapter dedicated to an exploratory review of Mary Shelley's well-known story about Frankenstein's monster), Winner, in the end, wants his readers to understand they have the freedom to "demystify" technology, even when it appears in monstrous proportions wielding unheard-of abilities. From his point of view, it seems reification of technology is a choice we make and one that we can reverse.

If Winner failed in that approach, that is, if simply opening our eyes to the fact that we created technology and can therefore "own it" is not enough to engender changes in our attitudes and actions toward it, it is perhaps because in the end, those beholden to a tower glittering in the sky cannot be asked—truly, cannot be expected—to take the time and effort to ask what the real benefits and consequences of that tower are for the inhabitants of Babel. While this statement no doubt sounds glib to many, we believe when we're done recounting the hundreds of deaths directly attributable to digital reliability issues (serious problems that occur in multiple fields and venues), and when we've also shown how trite our remediating responses to that lack of reliability have been, even technology's true believers may begin demanding more accountability from those "building the tower."[14]

Towers and Babble

Speaking of towers representing the actions of man and the value of *human* action and communication, I want to recount two incidents that I endured many years apart and in two different high-rise buildings. In the first incident, as a young college undergrad living on the sixth floor of a 12-story dormitory, I put my head on my pillow after a long evening of study (or more likely, after enjoying some amusing substitute for it) and glanced across the dorm room at my wind-up alarm clock. I had placed it a good ten feet from my bed, on the floor near the door, because the long hours of college life had taught me that an alarm too close at hand was too easily shut off and ignored. Having assured myself the alarm was set, and having just fallen into a deep sleep, the loud banging on my door at about 2 a.m. confused me. *That was quite the alarm*, I remember thinking in the first haze of awakening. Then I heard the unmistakable voice of my floor's Resident Advisor (RA) loudly explaining through my closed door that we had a fire and were to exit the building immediately.

I stumbled into the hallway and wondered immediately how I could have remained asleep at all. Students scrambled everywhere, alarm sirens wailed, and the whole floor filled rapidly with smoke. I knew there were two sets of stairs and an elevator in the building, so I headed toward the set of stairs that led directly to an outside entrance at the base of the tower. As I rounded the hallway, I came to the dorm's trash chute—a hinged metal door in the outer wall of the hallway that opened into a metal chute that went all the way to the base of the tower and emptied into a large dumpster in the basement. Black acrid smoke poured from the *closed* trash-chute door. Our floor's student chaplain stood guard by the smoking chute, stopping students from walking near it. I looked at the smoking chute and then at him. "You can't use this set of stairs," he said. "Use the other set."

As I instantly reversed direction and headed rapidly toward the other set of stairs, I looked back over my shoulder at the faces of the chaplain, the RA, and a couple other young men who carried flashlights and marshalled sleepy-eyed students to the only set of stairs available for a safe exit. There was indeed smoke everywhere, and as we hustled our way down the stairs, we wondered aloud where the source of the fire was and why the other set of stairs was impassable. Why was smoke pouring from the trash chute? Was it acting as a chimney for some horrific conflagration on one of the floors below?

The lower we went in the building, the more we could hear people coughing and choking in the echoing stairwell. All bets were off, and we had little or no situational awareness. For instance, I would have gone down the wrong set of stairs had the student chaplain not stood guard and forced me in the other direction. I remember his somber expression, and I could see a bit of fear in his face as he directed students to safety, but he stood his ground, flashlight in hand, making sure he guarded the unusable staircase and kept folks away from the fiercely smoking trash chute until the floor was cleared of all students.

As it turns out, the fire was in and around the dumpster in the tower's basement, and the set of stairs I would have chosen for my frantic exit would have taken me directly to the fire's source. We emptied into the frigid January night and watched the firetrucks and firemen hauling freezing hoses to the dorm's basement. Within a surprisingly short time, the fire was out, and then the firemen finished several hours of cleanup. The story ended well, except for the fact that none of my classes the next morning were cancelled. Not one serious injury, however, and a building saved with no serious injuries or deaths. Many of us noted the actions of the student RAs and chaplains and their roles in keeping us all safe. When the fire broke out, they had little more information than we did. Yet they "stayed with the ship" while black smoke poured into the hallway, and they conveyed the information and directed the actions we needed to reach safety quickly. I didn't give it enough thought at the time, but their unrehearsed actions were sensible, timely, and potentially self-sacrificing, and they certainly prevented injuries and may have even saved lives. They were in place, accepting responsibility and providing crucial information and face-to-face guidance at a critical point of human need. And they coordinated our timely and safe exit before anyone had even heard of a cellphone, texting, crowd sourcing, tweeting, or navigation apps.

Fast forward about 25 years, and a computer-consulting gig had me staying the night in a high-rise hotel in a major eastern city. The hotel was roughly the same height as my old college dorm, and this particular night I had a nice suite on the tenth floor, completely around the back side of the tower, as far away from the noisy elevator as you could get (yes!).

I had a presentation to give to a potential client the next day, and the combination of travel stress and the usual pre-presentation worries gave me a fitful night of sleep despite the lovely room and accommodations. Somewhere around 2 or 3 a.m., I found myself wide awake and decided to channel surf and dig around the hotel's free Internet. I did this for about 20 minutes, weary-eyed, until suddenly the TV died and every light source in the room went blank; the blower for the ever-running heating and air conditioning system went eerily silent as well. The pitch darkness was something I'd never experienced in a big city, those places that won't ever go silent or stop glowing when you most need peace and quiet. I sat for a few minutes in the darkness wondering how long a huge building that was part of a massive and famous hotel chain could stay without power. Not long, apparently, as the power came back on after about ten minutes, and the familiar hum of the HVAC returned. I decided then that I'd better locate my shoes and a few other items that had turned out to be impossible to find in the pitch black. I set my shoes, shirt, and pants near the door and decided to give sleep another try. The power was back on, the heating system fan was humming again, and all was well.

Five minutes later, while I lay in bed, every powered item in the room went off again. I looked at the base of the door, hoping for some light from the hallway. Surely a power outage in such a large modern building didn't wipe out everything? But, for the second time in the span of 15 minutes, I saw no light source of any kind. So I inched my way in the darkness to the door and opened it. Outside in the hallway, complete darkness, no movement of other guests, and no sound at all—except...

In the distance I could hear a voice of some kind. I couldn't make out what the voice was saying, but it didn't sound like it belonged in the hallway of an expensive hotel in the middle of the night. I strained my eyes in the pitch black and tried as best as I could to focus my hearing on the voice. I couldn't make out a word. But the voice babbled on about something.

At this point, I considered what could really be happening in the hotel. If the power was simply out for a short while, i.e., if that was the only issue in the hotel, a recorded voice would certainly be overkill in the middle of the night. That's the first time it occurred to me that maybe more was going on than just a popped circuit breaker. On the other hand, if it were a major emergency, wouldn't I see backup lighting, hear an alarm, see rushing hotel employees?

I returned to my room—I'd been careful not to lock myself out of the digital key-enabled door (would it even work during a power outage?)—and fumbled around for my shoes and jeans. Once dressed, I grabbed a few essential items and started feeling my way tentatively around the hallway toward the front of the round building. In the pitch black, as I drew closer to the elevator, I began to make out the sound of the recorded voice I'd been hearing.

Once I was close enough to understand the voice, what I heard seemed inconceivable given the circumstances, so I hope my lack of reasonable action, let alone an urgent response, can be pardoned. This droning, digitally recorded voice, unaccompanied by either an audio- or light-based alarm, was mentioning that "there may be" an issue related to fire in the building. I was to "avoid the elevator" and "please make my way" via the stairwell to the lobby exit.

The first time I heard the word "fire," I froze in my tracks to (finally) take full stock of the situation. The hotel was packed with guests—this much I knew. The power had gone out twice and was still out, and the outage seemed comprehensive on floor 10 of the hotel. No illuminated exit signs. No night lights. Not even a blinking light on a hallway courtesy phone, let alone the flashing fire warning signs I expected.

I sniffed the air. Yes, I thought maybe I was catching a whiff of smoke. Again, for a moment, I froze. Smoke in the air, no fire alarms other than the droning recorded voice that could only be heard within about 25 feet of the elevator, no emergency lights, and not one other guest or hotel employee in sight. Memories of 911 tower escape stories and TV footage came flooding to mind. In the middle of the night, with smoke in the air, one high-rise fire in my past, and not a single light, alarm, or person in sight—my imagination kicked into high gear.

I expected to round a corner and find fellow guests, helpful employees—maybe a fireman or two—someone maybe with a flashlight in the impressive darkness. But for more minutes than you can imagine, none of these revealed themselves. I did, with much stumbling, finally find the moderately lighted stairwell. I wasted no time starting down the heavy concrete steps, the droning and useless digital recording my only accompaniment for many flights of stairs. Not another human being in sight. Then the smoke started in earnest. Not choking, life-threatening smoke, thankfully, but smoke it was, visible and certainly full of electrical-fire smells.

After descending what seemed to be dozens of flights of stairs, I saw a young woman and her daughter in the stairwell.

"What's going on?" I asked.
"I don't know," the woman said.
No matter, I thought laconically. When they film me jumping from the flaming window, I won't be alone.

We clamored down the steps together, throwing out various ideas between our rapid, noisy footfalls in the concrete stairwell. I don't remember all the details of the hurried conversation, but I do remember the two topics of most interest to all three of us: Is the smoke going to get thicker? And, which door do we exit? Somewhere in there I may have unleashed a profane remark or two about the complete lack of hotel employees and/or guidance. (Sorry—I don't count barely intelligible digital recordings echoing in a concrete stairwell during an active hotel fire as guidance. Maybe it's just me.)

Finally, having loosely counted the number of flights we descended, but without so much as a hand-painted sign on a door to help us make a decision, we chose a door to exit the stairwell. Into some hotel kitchen we went. Again, not an employee in sight. We wound through a few stainless steel prep tables, then found someone who didn't speak the local dialect. Once again, the surreal; we're in a hotel fire and the immigrant kitchen worker apparently doesn't know about it.

Back out the door into the stairwell we went and down a couple more flights. I would have thought I was in the middle of a hungover dream if it weren't for the very real smoke.

Finally, we exited a door and landed in the huge marble-floored lobby, which was extremely dark, even for the middle of the night. Now there was no doubt about the reality of the fire, as fully dressed firemen hauled hoses and equipment into the lobby, while diesel-engine firetrucks droned with red lights flashing in the dark just outside the hotel's revolving glass doors. Again, I looked around and found absolutely no human guidance. I could have punched one of the fireman in the gut and run out the door like a character in a Bruce Willis movie, and I truly don't believe anything would have happened to me.

We exited the hotel and its constant flow of incoming fireman, weaving in and out among them like salmon swimming upstream, and I lost track of the young woman and her daughter. They were safely outside, though, so I set about the task of looking for someone I knew among the hundreds of hotel quests standing on the lawn and sidewalks in various states of partial dress.

I found my boss about a hundred feet from the lobby door. He was cool, as usual, not the least bit ruffled about our forced exit nor as concerned as I was about the complete lack of human guidance in the midst of a fire.

"Where's Bob?" I asked.
"I haven't seen him. What floor was he on?"

The conversation continued for a short while until it was clear to both of us that we did not know if our colleague Bob was safely out of the building, nor did we know if he was in imminent (or even eventual) danger.
Your mind moves quickly in an emergency—we all know that—but I'll own up to the fact that in the middle of the night, *my* mind doesn't always move quickly in the most rational direction; I found myself thinking for a brief moment, long before anyone knew the names Assange or Snowden, how ironic it was that so many government agencies could put their hands on personal details of our lives—where we traveled, where we stayed, with whom we spoke—and yet in the middle of a fire, my boss and I, both as digitally savvy as they come, had no idea where our friends and colleagues could be found.

I told my boss I was going into the hotel to report that Bob was apparently missing. I didn't wait for his approval. Sorry to invoke a scene from a Hollywood movie for a second time, but I fell right into line with some fireman lugging equipment into the lobby, half expecting someone to stop this un-uniformed guy from walking into a building on fire, and just trudged right in the front door. Not one fireman said a word to me, and at that point I realized that all Hollywood movies are true.

In the darkness of the lobby, I could make out a dimly lit concierge desk. The concierge himself was long gone, but a uniformed female security guard stood her ground behind the concierge's podium.

"My friend is somewhere in this building, and we don't know where," I offered with no introduction and little elaboration. After all, this was an emergency; I talked quickly because I was sure I wasn't supposed to be

inside the burning building and would be promptly dispatched back outside. I gave her Bob's full name and asked if she could ring his room to make sure he was safely out. She looked up slowly, almost like a harried retail clerk. I'm not sure exactly what I anticipated her to do or say, but her response will be remembered as one of the least expected of my life. She said slowly, without the slightest trace of urgency or irony, "Well, we're not supposed to use the phones now."
?!..&^*%o!! ^& !!!!

Clearly at that point in my life—post 911 but not yet into the wonder years we're enduring now—I had not adjusted myself to life among the silicon-enabled and digitally enlightened. I actually tried again to explain to her that in this massive hotel that was on fire somewhere, I had a friend and colleague who apparently had not awakened to the droning digital recording that was the hotel's excuse for a life-saving fire alarm.

No comprehensible response came from the guard. So, one more time, back out the door against the flow of firemen I went. I once again found my boss standing in the crowd and told him the incredible story.

> "Do we know what room he's in?" my boss asked.
> Of course not. I hadn't been able to gain even that bit of information from the guard.
> "Do you have a cellphone?" I asked.
> "That's a good idea."

We had no idea what room Bob was in, but my boss did know Bob's *cellphone* number. Digital technology to the rescue. He gave it a try. Then another, and another. Bob did not answer. As with so many things digital, we didn't know if the lack of an answer meant his phone was not working or ... well, there were many other reasons someone might be unreachable in a fire.

Eventually, I gave up trying to save Bob. What a thing to do, really. And yet, the entire inter-connected digitally wired enterprise was all around us, letting us know with each worry we expressed that all was well. If not "well," then at least hopeless (a word that seems to be partially re-defined in our digital environment as meaning "in a state in which human action would be pointless"). Lack of hope in the digital age means you do nothing—or at least you make no waves.

I thought back briefly to the *first* high-rise fire I'd endured and recognized how radically different everyone conducted themselves in that fire as opposed to this one. I looked up at the tower, just as I had done a quarter-century prior in college, and wondered if we'd eventually see flames, wondered if the fire was small? Large? Growing? Under control? I even wondered if I'd have to watch Bob on a ledge at some point, deciding between burning alive or plunging to his death from the—well, I didn't know which floor he'd jump from. But it wasn't a pleasant situation knowing that neither the hotel employees, the firemen, nor the hotel's digital warning system had the slightest concern for my colleague Bob, nor any idea where he was or even where to look for him.

I'll end this story by saying that, thankfully, no one jumped from a ledge that night, and indeed the fire was contained for the most part to a large kitchen on the ground floor where an electrical short had kicked it off. We never learned from the hotel staff why the power in the whole hotel went off (twice), nor did we receive any explanation as to why the staff did not engage in helping the hotel guests get to safety. To this day, the only explanation I have is the one I've concocted myself: *that over-reliance upon the hotel's digital fire-warning system had led to unimaginable inaction on the part of the staff in regard to guest safety.*

To put this tragedy of inaction in perspective, had I not had insomnia that night, I would have slumbered as the smoke from the kitchen fire climbed the stairwell, and neither I nor Bob (who, completely unbeknownst to me, was in the room next door to me on the 10th floor!) would have left the building had the fire spread. People do die in high-rise fires. And people who administer buildings open to the public certainly have a *human* responsibility to ensure, to the best of their ability, the safety of those who are their guests in those buildings. All the digital warnings in the world and all the cellphones on the east coast don't change that interpersonal responsibility to one another. *Interpersonal responsibility to one another.* That's a phrase worth repeating, if for no other reason than the way we use digital technology today provides us with a layer of actions that appears to divest our human responsibilities. Appearances, indeed, can be deceiving.

Not many folks can say they've escaped two high-rise fires in the middle of the night. As with most intense emergencies, my memories of those events are strong, and the available lessons learned have remained with me over the years. Those two fires were so similar in multiple regards: they both started in the middle of the night on the bottom floor of towers that were approximately 150 feet tall; they both sent smoke up the only available exit ways; they both triggered the evacuation of the towers; and both were contained fairly rapidly by the actions of fast-acting fire departments.

The most notable difference between those two events—similar in so many other ways—was in the actions of the staff charged with the safety of the tower residents. And the significant fact that during the second fire, in our digital era, the evacuation of the tower was incomplete.

Hyperbole is hard to recognize and even harder to define when looked for within the age in which one lives; this is, perhaps, self-evident. However, departures from virtuous action occasionally emerge that make hyperbole, well, quite acceptable. I can say that the actions of the hotel staff that night in that DC hotel were abominable. Whatever training they'd been given, whatever digital tools, networks, and communications devices were at their disposal, and whatever the legal staff of this multi-billion-dollar hotel chain had advised them in advance, they acted wrongly, bordering on maliciously. When a building for which you are responsible catches fire while hundreds of your guests are sleeping on multiple floors, you must act to ensure they are led to safety while the fire is brought under control. If you do anything less than that for any reason, you have failed basic human responsibility.

As to what that statement has to do with digital reliability, we hope the following chapters provide some enlightenment and answers. We drive home the point that "digital reliability" means answering far more complex and impactful questions than "Did my cellphone connect?" or "Did the digital thermostat shut off the heater at the right time?" Simply the fact that many folks, including the proverbial movers and shakers in our society, believe that digital technology does indeed mimic the mind of man (or that it will do so in the near future) impacts how we choose to use digital technology; it also impacts a myriad of questions such as "Where is digital technology appropriate?", "What traditionally human actions and decision-making processes can we entrust to computers?", and "How many of our own responsibilities are we now free to ignore?" Ethical questions that are even more complex arise *after* we choose to abdicate action and responsibility to computers only to subsequently watch some digital system or device fail to the detriment of a fellow human being's health and wellbeing. In the example above, I wonder if the security guard who claimed she was instructed not to use the phone system ever wrestles with nightmares involving what might have happened had the fire expanded while my friend Bob slept?

In fairness to the ethically catatonic hotel security guard, she made the decisions to withhold both action and potentially life-saving information not because she was a psychopath or pyromaniac; she had become, as John Taylor Gatto and other keen observers of our times have described, a bureaucrat in uniform.[15] Digital technology in and of itself may not turn us *directly* into non-acting bureaucrats, but our collective choices in how we implement technology and, even more importantly, how we gauge, monitor, and mediate the full range of its effects on us can and do turn us into non-actors at some of the most unfortunate and dangerous junctures of our lives. We hope the chapters that follow will shed light on this fact and perhaps help us to change.

3 DEFINING DIGITAL RELIABILITY

Knowing When Something Is Amiss

Lived experience is a seamless web, but academia in particular encourages specialists to indulge in reductionist interpretation. ... It is not adequate to suggest that what shapes technology is science, since science is also socially shaped, and technology also influences science.[16]

— *The Shaping of Technology*, Donald MacKenzie and Judy Wajcman, eds.

Rather than being 'outside' society, technology is an inextricable part of it.[17]

— "Technological or Media Determinism", Daniel Chandler

I live and drive in mountainous areas, and as the price of gasoline continues to rise out of pace with my income, I've searched for ways to conserve. Riding a motorcycle helps save a few dollars, but only during the three days of the year it's not raining or snowing in my neighborhood. One day while on the road to the grocery store, I discovered I could turn my car's engine off and coast downhill without burning so much as a drop of gas for distances as great as three miles. My car has a manual clutch and transmission, so restarting the car at the base of the hill was a simple matter of "popping" the clutch and hitting the gas. Using this method, I didn't even have to put wear and tear on my car's starting motor.

I noticed one problem: the car ran horribly for 20 minutes or so after I used this restart procedure, banging and knocking and making all the sounds of a poorly running engine. My mechanic later explained to me that my method for "saving gas"—i.e., coast for a few miles and pop the clutch to restart my engine—would have worked wonderfully with many cars manufactured right up to the 1970s. However, with my newer car, I was both failing to save gas and simultaneously causing undue wear and tear for my engine. He told me my engine's computer wasn't getting its mandatory "reset" when I started it by popping the clutch instead of turning the key to the start position. The end of that story is that in no time at all I was paying hundreds of dollars for a new engine. The handful of dollars I'd saved coasting downhill through the mountains was long gone, and from my point of view, I made this somewhat costly mistake, which eventually destroyed a perfectly good engine, simply for the lack of awareness regarding the digital technology in my car.

Digital technology's often silent pervasiveness actually makes defining digital reliability in its broadest and most accurate sense—i.e., in the sense proportionate with digital technology's actual roles in our lives—a bit difficult. We have digital technology in devices and in places most people don't even consider. Those who run our markets and our government thrust digital technology into multiple areas of our daily lives; in some cases, we can see the changes and have known about them for years. Sometimes the changes bring good results, but certainly not in every case. Many years ago, I bought my first "digital" toaster from a kitchen appliance store where it was sitting behind a display card bragging that the toaster contained a "computer chip." When that toaster broke down faster than any kitchen appliance I'd ever owned, I made a quick switch back to the "chipless" variety—the exact kind I'd watched my grandfather use and occasionally repair for decades using tools no more exotic than screw drivers and pliers. We found that our toasted bread tasted identical using either machine, and the added reliability and ease of repair made the non-digital toaster the winner. As Intel's Brian David Johnson put it: "I think technology for technologies' sake is sort of silly. Do you want to turn your toaster into a computer? Well, maybe. But how is it gonna make your life better?"[18]

Other examples of invisible digital technology from markedly different venues abound, each a variation of cautionary tales. A simple trip to the local WalMart might, for example, mean walking out of the store with newly purchased clothing and other items embedded with RFID (radio frequency identification) chips.[19] These tiny chips, which use integrated circuits connected to antennae to transmit information about the item they're attached to, including its location—may seem simple, innocent, and unobtrusive. They are, nevertheless, a very sophisticated form of digital technology that brings both powerful benefits and unintended possibilities for misuse.[20]

Some digital technology is easy to spot. If it beeps or glows (or both) then it is usually safe to assume there are digital components involved. However, we use many products daily and intimately in which the digital technology is quite concealed; without digging further, we have no way to know digital technology is present. This is so, in part, because we have no laws or traditions that ask manufacturers, industries, or governments to notify us, the end users, of these products and the presence of digital technology. Not surprisingly, we find ourselves consistently unaware of digital technology's ever-expanding presence (or, far worse, unaware of the implications and consequences of our ignorance).

Our hope is that after reading this book, readers will deem as reasonable the time and effort required to explore where and when digital technology is present in our lives, and even, perhaps, choose to ask the important questions about the long-term implications of digital technology—the questions that begin with "Why?" and "To what ultimate end and purpose?" and "How will this impact people and society in the long term?"

The Robot Will See You Now!

Sometimes we find examples of less than fully understood digital technologies that are a bit more eye catching than a failed toaster or car engine. Robotic surgery is not exactly new, but its recent popularity in hospitals around the US has been phenomenal to watch. Billboards from multiple hospitals in my area show pictures of happy patients standing in front of robotic surgeons, beckoning more patients to give this latest triumph of digital technology a try. What, however, do we really know or understand about robotic surgery from a billboard? I've passed these advertisements many times and have considered asking the many medical personnel I know how the robots are making inroads and if those inroads are penetrating various hospitals and medical specialties evenly.

Recently, Lindsey Tanner, medical writer for the Associated Press, provided some details of how robotic surgery is fairing—details that really should, from at least one perspective, be flowing freely in our press (and there's yet another of our themes);[21] *that is, it's fair to ask whether or not the decision to allow digitally controlled machines to perform the most intimate possible physical procedures on our family members, friends, and neighbors has been properly vetted,* and whether the motivations, economic factors, and administrative machinery bringing these robotic surgeons into our lives are appropriate. Are we constantly monitoring these factors so that they *stay* appropriate, given the seriousness of such a change for our society?

Tanner asks: "Is it time to curb the robot enthusiasm? Some doctors say yes, concerned that the 'wow' factor and heavy marketing are behind the boost in use. They argue that there is not enough robust research showing that robotic surgery is at least as good or better than conventional surgeries." She goes on to describe several reported instances of problems and failures during robotic surgery. A few descriptions, while very serious (or, much more rarely, *deadly*) for the patients involved, sound no more alarming nor significantly different than those we hear about from conventional surgery—a nicked blood vessel during a hysterectomy and a death during spleen surgery (and we know that occasional deaths certainly happen during even routine non-robotic surgeries). Nonetheless, a few other stories grab our attention in a different way—a robot surgeon that inexplicably struck a woman in the face during surgery and another that grasped tissue during colorectal surgery and would not let go until it was forcibly shut down.

While some stories like those have headline-grabbing potential, the real story, we believe, is summarized by Tanner's quote of Dr. Martin Makary, a Johns Hopkins surgeon who co-wrote a paper about robotic surgery. He claims problems linked with robotic surgery are underreported, including those with "catastrophic complications." Even more significantly, he notes: "The rapid adoption of robotic surgery ... has been done by and large without the proper evaluation."[22]

And there's a theme we'll encounter over and over again: digital technology put into place—without discussion, without argument, without so much as a talk-show examination of its value, inherent risk, or effectiveness, let alone a serious discussion about potentially unforeseeable social, political, or cultural impacts.

We are not the ones to suggest, for instance, that the nation needs a "Digital Impact Review Board" or any other such convention. The belief that a small group of "experts" can always, without fail, find rational arguments for or against a given solution is often at the core of the thinking of a control-obsessed people, after all. But we do note along the way that those who embrace unremitting rational control over society—including the professionals, the amateurs, and social engineers—with the greatest irony, if not hypocrisy, make a seemingly endless series of non-decisions by rolling out tools and infrastructures with massive implications for all of us.

This fact reinforces our core argument that the *real* definition of digital reliability—that is the degree to which we can trust digital technology to help us rather than hurt, debilitate, or *enslave* us—is under-considered from coast to coast, from pulpit to printing press. As Stephen L. Talbott states so pointedly:

> 'But surely,' you may say, 'all our technological advances do represent an accumulating gain. Nearly everyone agrees that technically mediated services are getting better!' But this confusion of technical advance with human benefit is the heart of the reigning lie.[23]

So where do we go from here? Do we list more articles like Tanner's, laying out a simple "sky is falling" argument—an argument that would certainly have potential to play a part in the thesis, antithesis, synthesis approach we all know so well?[24] Along the way, we will indeed describe digital failures with gut-wrenching results. The emotion inherent in these accounts is not the point, however. Neither are the statistics we can mention (or that readers can derive) from these anecdotes. We do wish, however, to make the point that digital reliability is both a huge factor in our daily lives and a factor that is routinely ignored—sometimes almost appearing to be systematically neglected even in the aftermath of deep-impact failures of digital technology. As Jianto Pan of Carnegie Mellon University (CMU) notes in his paper "Software Reliability": "With processors and software permeating safety critical embedded world [sic], the reliability of software is simply a matter of life and death. Are we embedding potential disasters while we embed software into systems?"[25] If this sounds like conspiracy thinking, so be it. That is neither our intent nor our mindset—simply a Cassandra-like quotation flowing from CMU—one of the most significant foci of digital technology in the world.

Definitions—Or How We'll Build Them

For all its potential complexity and for the incredible diversity of its applications, digital technology, for our purposes, surprisingly still breaks down into two general categories: software and hardware. These categories certainly won't remain accurate or be all-inclusive of digital technology for long, because we already have at least one entire industry (bio-genetics) integrating digital technology into biological entities. Brilliant scientists from that industry are experimenting with storing text (for example, a book) using DNA; but we can hardly say that using human DNA to store information is manipulation of hardware, and we're not yet ready to concede that DNA is "software" in our traditional understanding of that word. Yet the act of storing information in DNA is most certainly a form of digital technology.[26] For our purposes, though, we'll analyze various industries' use of digital technology using digital technology's two traditional categories—software and hardware.

Pre-defined and unambiguous definitions of reliability (or the lack thereof) can also be tricky and subject to much debate. Software engineers sometimes use very specific definitions of reliability—definitions that we can confirm and track tidily through statistical analysis.[27] And very likely some readers expect a book about digital reliability to base some portion of its findings on statistical analysis. We do not eschew statistics completely in the chapters that follow, nor do we analyze digital reliability from a single specific point of view, such as theological, anthropological, behaviorist, etc. (We remind ourselves that praying drafts of this book don't get lost in laptop crashes doesn't count towards a theological foundation.)

One of our theses here, though, is that a purely statistical approach to tracking digital reliability is missing the mark—in many, many ways. Statistics notoriously overlook the human element, certainly when those statistics are partially or carelessly employed. (Four of five tech writers surveyed—and their avatars—agree with that last statement.) And again, with the seemingly ubiquitous quest for cyber technology flourishing around the world, the human element will continue to rub elbows—and increasingly, other body parts—with digital technology. That said, our definitions require a modicum of discussion.

Starting Point for Definitions: Why We Should Analyze Digital Reliability

(Or—A Tale of Two Cockpits on the Worst of Nights)

All of the aircraft I flew during a career in aviation used digital technology to some degree—some barely at all, but others almost exclusively so. One particular night, I flew through the Rocky Mountains in a cockpit crammed with glowing digital paraphernalia. We had so much electronic equipment in that plane that the modifications the mechanics and engineers made over the years had actually affected the weight and balance of the aircraft. I wasn't complaining, however, as we were flying in the mountains on a cloudy night, and despite the limited visibility outside, all of our wonderful navigation aids and visual displays helped us know exactly where we were at all times. With our advanced equipment, we even knew the whereabouts of other aircraft in the area—some of which were piloted by men just as balmy as us—at low levels and high speeds in the dark mountains.

As I looked out into the night, occasional lights on the desert fringes slid by beneath wisps of cloud. I thought back to my very first night flight in which a pilot friend and I leaned our heads against the windows, working to keep sight of Route 30 as we wound our way out of Pittsburgh into the mountains—in a small plane built while television was in its infancy. At that point, I hadn't yet gotten my private pilot's license, so I was fascinated by every aspect of the adventure, no matter how old and small our plane was. The cars and their headlights wandering through the wavy hills of western Pennsylvania to my hometown in Westmoreland county provided wonderful ad hoc navigational aids during that first night flight. And, having seen my hometown from the air for the very first time, when we turned around to head back to Allegheny Airport, a simple tune of one of our radios to KDKA's radio tower frequency (AM 1020) gave us a needle in a dial that pointed the way back to the 'Burgh. All was well on that first flight—until the lights went out.

A simple plane like the one we flew on that first night flight doesn't hold much sophisticated digital equipment, so we weren't in a big jam when the lighting failed. But we did need to know at least a couple things in order to get back on the ground safely after our cockpit went dark. Airspeed is critical for any pilot in any airborne situation, and that night when the cockpit went black, we couldn't even see the plane's airspeed gauge. My friend Jim calmly reached into his flight kit and pulled out a small flashlight. "Here," he said. "Turn this on and cover about half the lens—so you don't blind me—and point it at the airspeed gauge."

I wasn't about to argue—just kindly tell me, Jim, which one is the airspeed gauge? As to navigation, we had that covered thanks to good familiarity with many ground references—which helped us find our way even on a dark night.

Holding a flashlight on a round dial in a dark airplane wasn't as fascinating or enjoyable as watching cars on Route 30 wend their way through the hills in the night, but I wanted back on the ground at some point, and I was counting on Jim being of the same inclination. So I cooperated, doing my small but important part, and Jim expertly parked the plane on Allegheny's runway while I held the light on the gauge the whole way down final approach.

Fast forward to the night in the Rockies, and the navigation aids in our plane differed immensely in at least one respect: I needed no automobile headlights, winding roads, or radio towers to help me out. The array of screens and displays at our disposal was staggering and reassuring in the night. When you're cruising along at high speeds and low altitude and you have redundant systems all over your cockpit giving you visuals and statistics about obstacles, altitudes, television towers, smokestacks, other aircraft, and even weather, it's truly impressive.

But not nearly as impressive as the stunning pitch black that enveloped the cockpit about an hour into the flight. Every screen, every dial, and every gauge and light in the cockpit went dark, and did so without a moment's notice. In a couple thousand hours of flight time, this was only the second time I'd seen a cockpit go dark, but here I was again.

This time, as a seasoned pilot, I knew that equipment failures such as these happen, and they happen even in aircraft and machines equipped with triple-redundant systems—as was our aircraft. I also knew that I was supposed to maintain **situational awareness** at all times, and that would help me keep the aircraft where it should be as long as we maintained airspeed and controllability—both of which we had. During this particular emergency, with a good engine and fully capable flight controls, the fact that in one second we went from having a lighted electronic cockpit that made us feel like we were flying a "mini AWACs" (as my copilot used to call our plane) to flying a pile of blind steel on a cloudy night at low level in high mountainous terrain—this dramatic shift of events was *supposed* to mean next to nothing to me. How I *really* felt that night is a different matter. Nonetheless, I immediately did what we'd been trained to do to survive the first important minutes of a low-level emergency: I started a climb. Next to me my copilot obligingly opened his paper map (yes, we still had one on board!) and shined his red-lensed flashlight on it.

"Climb to five thousand feet," he told me calmly.

What? Five thousand feet when we're amongst mountains topping out at over twice that height? I didn't want to question him too much—he was my friend and he was a skilled navigator—but I felt like we were headed to the steerage section of a sinking ship if we climbed to only five thousand feet.

After a brief discussion about the veracity of my copilot's paper map, we safely leveled at five thousand and turned to the task of contacting the tower to let them know we had an emergency. I didn't use the magic "mayday" word—which I only used once in my aviation career—but I sure *thought* about using it. Nevertheless, when the tower heard we had complete lighting failure in the cockpit, we got attention fast. They let us know they were going to arrange for an escort plane on final, making sure we'd have no trouble finding the runway in the cloudy night.

At about this point in the emergency, when he had stabilized at a safe altitude and been in contact with folks who had us on radar and were poised to help, I was finally able to take full stock of our situation. We realized the lights and power went out in the cockpit for a reason. Trouble was, none of our indicators were letting us know exactly what that reason was; and about a thousand hours prior in my aviation life in a different aircraft, I'd had an electrical fire on board—resulting in an emergency landing in a field; the aircraft we were flying this time wouldn't fare so well in a field landing, to say the least. So there we were, my copilot and me, flashlights in hand, trying earnestly to figure out how so many independent circuits could all go out at the same time in a plane that was wired so redundantly. In short, we were busy trying to figure out if the light failure was a precursor to far worse problems. This made a typically routine RTB (return to base) far more taxing than normal.

After the tower finished arranging the launch of our escort plane, the radio cackled again, and they told me to do a "fix-to-fix" navigation maneuver to a point just in front of the runway. A fix-to-fix maneuver—bane of student instrument pilots everywhere but done routinely a thousand times a day all across the nation by experienced pilots—was supposed to be easy. Trouble was, to execute that maneuver, you had to be able to see some of your instruments. And we couldn't. I was far too proud to tell my copilot or the tower that what I really wanted was a snap vector from radar to tell me where to point the nose of the plane. That's what I should have done, given my concerns about larger safety issues. Instead, I struggled with a flashlight and some paltry backup lighting my copilot rigged (*ala* Apollo 13), and we completed the fix-to-fix maneuver just about the time our escort plane showed up on my wing.

This story also ended well. The escort plane made life down final approach easy, even though it scrubbed my chances of writing up the emergency with me as the story's hero. And our own plane, dark and dumb as a Wright Flyer, kept its engines and flight controls running perfectly all the way to the runway.

Professional pilots are trained to learn from each flight experience and certainly from each flight crisis, i.e., from both our own emergencies and those of fellow pilots. Once on the ground, we debriefed our flight in an attempt to gather some "lessons learned." I never did let on to my copilot just how lost I felt when all those screens in the cockpit went blank; but the fact is I had so little situational awareness at the initial moment of cockpit darkness that it's hard to say what I would have done without him beside me with his flashlight in hand and his nose glued to the map. Without question, I had become a different pilot than the one who did his first cross-country solo flight over the fog-covered Okefenokee swamp with few navigational aids other than a map, a watch, and a compass. In the interim, I'd been introduced to digital flight technology that simultaneously provided me with new abilities and quashed some I'd developed previously. Digital technology *gave me* "*eyes*" in the night that no human eyes could match and *took away* elements of my inherent **situational awareness**—a type of awareness that needs regular care and feeding (through constant practice) in order to be effective at times of emergencies.[28]

As I reflected on my own less-than-stellar performance in the emergency, my thoughts went back to some recent aviation training I'd had. Unlike many of my peers, I *liked* spending time in our high-end digital flight simulators. I booked more time in those things than just about anyone else I knew. Our digital simulators, housed in a large windowless warehouse-type building, provided amazing practice possibilities. We could simulate just about any emergencies the real aircraft might encounter, and we could conduct as many long-range instrument and "night flights" as we wished, at a fraction of the price of a real flight. The only flight situations we couldn't practice in earnest were the age-old, tried-and-true flights in which we relied upon visual reference to the ground and skyline for honing our visual navigation skills and building up our general situational awareness. That's because the visual depiction our simulators provided was mostly limited to computer-generated "night" conditions shown in TV screens that rendered the view through the front of our cockpit. Side and rear visibility remained un-replicated in our simulators. But given the simulator's array of the real aircraft's instruments and flight controls at our disposal, most of us thought the lack of side and rear visual representation was a relatively small omission.

Yet, situational awareness about where we were and the awareness of visual navigation aids and references I might use to keep our souls attached to our bodies were probably the factors most clearly missing in my piloting responses the night the lights went out in the Rockies. I realized that unlike my years spent in much lower-tech aircraft—in which visual reference with the ground, night or day, was nearly always obligatory for safe navigation—flight in a high-speed, low-level fixed wing aircraft had taught me to rely upon radar, sophisticated digital navigation aids, and systems encompassing a host of other electronic and digital devices that removed the burden of maintaining visual situational awareness from my piloting shoulders. Without question, I had inadvertently developed patterns of over-reliance on instrument-flight techniques by spending more time in the simulators than I should have. In my own defense, I'll note that no one *ever* indicated that too much time spent in the simulator might lead to bad habits or an over-emphasis on one piloting skillset to the detriment of other vital skillsets.

My helplessness that night in the mountains when the lights went out showed me clearly that the digital navigation aids in the cockpit were a mixed blessing. When they were working, my life as a pilot was easy; I relied upon them and my copilot to tell me what was happening literally in every direction and dimension. When they failed, the skillsets I would have otherwise kept "in shape"—dead reckoning, map reading, interpretation of external visual cues on the ground and the horizon, and even maintaining my "night vision"— languished.[29]

The emergency event we experienced was drama enough for two pilots flying in the night, but by no means was it earth-shaking headline news. In the end, I suffered only from a nicked pilot's ego (that needed a bit of bruising anyway) and a bit of added stress as we limped our way back to the field without my usual digital crutches.

Yet this incident illustrates one of our key points about digital reliability: we need to measure the digital wonders showing up continually in our lives not simply by their abilities and the tasks they perform for us, but also, just as importantly—every bit as necessarily—we need to measure and consider what human and social tasks, abilities, traditions, skillsets, and opportunities they displace. In saying this, we're not positing a variation of *Arbeit macht frei,* nor is this a quasi-Marxist call for less "alienation of the worker from the work of his hand."

In this book, we discuss many problems and issues (from many fields of endeavor) related to technology-aided ambiguation of roles—such as I experienced in the dark cockpit in the Rockies.[30] We encounter practical issues, events, failures, and even real-life tragedies that have far less to do with abstract concepts and political theories than they do with real-world, in-your-face human failures. In many cases, these failures link back to decisions (and *non*-decisions) to replace human action, decision-making, and socialization with digital technology.

I did say "human failures," recognizing that readers can ask, fairly enough, whether a simple solution might be for all of us to strive to adapt better to the now-ubiquitous presence of digital technology in our lives. One of the key questions we hope to address in this book is: "Is it incumbent upon us to accept and adapt to digitally-engineered task and role displacement— wherever and whenever we encounter it?" We hope to answer that question—or better still, to help readers answer it for themselves—in the pages that follow.

4 TECHNOLOGY-INFLUENCED AMBIGUATION OF ROLES

We're tempted to label the main issue we discuss in this chapter, which is *ambiguation of roles*, as "cognitive forces." But this book preaches the dangers of careless reification, so we'll avoid the borderline hypocrisy of granting "force" status to a human concept. Note that any of the topics we discuss in Part One of this book could probably occupy a book in and of itself. We cover certain fertile topics here in Part One because they form (or name or help to describe) a social fabric in which we engage (or, more importantly, *fail* to engage) certain types of discussion about digital technology. Far more importantly—and this is why these issues are insidious and perhaps even inherently dangerous—the topics we discuss in Part One too often *disable and/or distract* discussion about digital technology and digital reliability, discussions that we should, without any doubt, engage in collectively and continuously.

The list of topics in the chapters that follow is not complete, but we believe it covers many of the issues those of us who are "insiders" in digital technology encounter regularly. Fellow writers, such as Postman, Gatto, Winner, Morozov, and Talbott, cover several of these topics under different names. In some cases, our discussions provide overlap with concepts another cyber-skeptic writer has labeled differently. Readers who, for example, understand Morozov's thoughtful discussions about "solutionism" and compare his thinking to our concerns below about ambiguation of roles and Hegelianism will note much overlap; we cover similar topics that are simply wrapped in different labels.[31]

In this chapter, we begin tackling our list of concerns with a look at technology-influenced ambiguation of roles, and we cover additional, related, issues in chapters 5 and 7. In Part Two of this book, we focus on digital reliability as it impacts various industry sectors. The concepts covered here will remain mostly a tacit part of our discussions in Part Two. In other words, we lay our conceptual foundation for issues with digital reliability here in Part One, and then we explore specific failures of digital technology in the second part of the book.

Why Consider Technology-influenced Ambiguation of Roles?

Digital technology grows in pervasiveness as more and more of us embrace its tools and toys; product- and person-embedded digital technology appears to be on a trend toward ubiquity that will make the Internet's current role quaint by comparison.

As Evgeny Morozov notes in *To Save Everything, Click Here*, many try to say that we are re-living an era similar to that of the 15th century AD and the advent of the printing press—with the difference being that we can stand back and see what is happening as digital technology expands and its influence explodes, self-reflectively watching with a type of collective Internet-enabled "meta-cognition."[32] From the extremes of this viewpoint, we can find folks who have no further need for the field of anthropology; the Internet provides such immediate and pervasive connectivity and feedback that we each understand the whole of world events and global mentalities (is there such a thing?) in "real-time," perhaps even precluding the need for further formal study of culture, societies, mores, etc. Take this a step further—perhaps only a half-step further—and we can project that for some thinkers and pundits, the Internet potentially brings the end of theology, sociology, anthropology, and perhaps even the 20th century's redheaded stepchild, psychology.[33] One of the many dangers of this sort of thinking is related to the age-old concern of some sociologists and anthropologists that the study of a group influences the behaviors of that group, no matter what precautions those doing the studying put in place.[34] And few technologies have ever created more of a sense of "being observed" than the Internet. A related problem is that experts in those fields have always had difficulty studying or fully understanding the "outliers" of a group—those who would refuse to be observed or who, upon recognizing the presence of observers, opt to remove themselves from study (or otherwise alter their "normal" behaviors).

To that point, there are many brilliant folks who choose not to embrace digital technology in their lives—writers, thinkers, philosophers, and plain old everyday workers. For those in our media and government who refuse for various reasons and motivations to include these outliers in their thinking and planning, the needs, wants (e.g., privacy), and potential contributions of these people increasingly go unattended. The narcissism (bordering on solipsism) of those who trumpet a new era of the Internet and who embrace *whatever* form that may take has hurt and will continue to harm those who simply want to live their lives outside the potentially smothering embrace of machine-enabled communication, commerce, and socialization.[35]

As Morozov explains clearly, the Internet is not enabling limitless self-examination, certainly not in any honestly new way. People are no closer today to self-understanding—let alone collective self-understanding—than they have ever been. The new deadly danger, though, is that we now set aside any semblance of humility on the issue of self-understanding, believing that the wonders of social media (or virtual worlds or real-time voice and image communication—name your "wonder" digital technology du jour)—are providing us with a deeper understanding of our neighbors, both near and far, than they are. The depth of that deception—i.e., the belief that digital technologies have shrunk the social and anthropological mysteries of the world in a genuine wisdom-inducing way—is far, far greater and potentially damaging than any of us can measure or predict.

At this point, Morozov would probably note that even this danger itself is not new, as many eras, or epochs as he would refer to them, before our own have mistakenly believed some new science or approach to viewing the world suddenly and completely made all things new. One of the most recent of those epochs, The Enlightenment, even takes a name that trumpets the self-congratulatory attitudes of those documenting, and presumably embracing, its tenets.

The trends, dangers, and possibilities we'd like to analyze, though, go beyond the Internet and those who delude themselves about its innate ability to make all things new. While the Internet and digital technology *are* new, and while we may use them to do and create wonderful new things and to develop novel activities and actions, some warn that if we forget that digital technology originated with us, with human beings, the danger to ourselves is no less than a self-inflicted reduction to automatons.[36]

The human pain, displacement, and confusion caused by technology-influenced ambiguation of roles is reason enough to analyze all things digital with a fresh eye. We believe that such an analysis of digital technology should build upon a new and vastly expanded definition of digital reliability—a new definition that goes beyond statistical analysis of computer task failures and includes analysis of how, where, and when digital technology causes human pain or displacement. And in using the word "pain," we include spiritual, moral, individual, societal, and even professional pain.

Sherry Turkle, MIT colleague of Joseph Weizenbaum and heir apparent to his deeply rooted skepticism about digital replication of human skills and abilities, encountered many examples of role confusion and social displacement in her decades of research into digital technology. In her most recent book on the subject, she said this about her findings in robotics research (a subset of digital technology currently garnering powerful support and attention around the world):

> Winston Churchill said 'We shape our buildings and then they shape us.' We make our technologies, and they, in turn, shape us. *So of every technology we must ask, Does it serve our human purposes?* -- a question that causes us to reconsider what these purposes are. ...
>
> It is one thing to design a robot for an instrumental purpose: to search for explosives in a war zone or, in a more homely register, to vacuum floors and wash dishes. ... What are we thinking about when we are thinking about robots? We are thinking about the meaning of being alive, about the nature of attachment, about what makes a person. ... What are we willing to give up when we turn to robots rather than humans? To ask these questions is not to put robots down or deny that they are engineering marvels. *It is only to put them in their place.*[37] [Emphasis mine.]

In short, Turkle, who is a psychoanalytically trained sociologist and psychologist, began her many years of research into the upper echelons of digital technology (i.e., artificial intelligence) as a "tech believer." She wrote early books filled with familiar praise of networked computers and the "advances" she, along with thousands of other writers, projected for all of us. She even describes initially resisting the highly regarded Weizenbaum's own skepticism about digital technology when they taught together at MIT in the 1970s. As she admits in her latest work, however, she did not then see what early developments in digital technology "augured."[38] The horrific anecdotes she now relates from her research make some of the late Joseph Weizenbaum's own churlish cautions about the dangers of over-embracing and over-estimating the benefits of digital technology seem, by comparison, like mild and under-informed fretting.

Where and how does ambiguation of roles begin? That is, how do we start with amoral "fun" machines that seem to augment or imitate familiar human actions and activities and end up with extreme confusion about human roles in relation to digital technology, a confusion worthy of multiple best-selling books? Even as we write this book, the entire world is abuzz with concern that the digital technology used by the US and UK governments to "keep the world safe" has instead turned the globe into—literally—an electronic prison. This is perhaps a form of *meta* ambiguation of roles; patriotic, would-be peacekeepers began with a well-thought-out acceptance of a clear-cut ambiguation of roles—i.e., they decided it was "okay" to begin using technology (in lieu of humans) to watch and control potential bad guys by invading the privacy of *some* communication of *some* key world figures. Along the way, they seemed to have allowed the "objective" use of machines for purposes of gathering and analyzing our every conceivable communication (and even our thoughts) to blur the initial moral decisions and issues involved. After all, signals intelligence is a type of digital technology use that seems particularly susceptible to generating a false sense of objectivity. Spies invading others' privacy with electronic devices—and doing so remotely and without the slightest hint of restraint—sometimes act as if they believe they're firmly on the same moral ground as medical surgeons cutting into the deepest parts of another human being: The end always justifies the means, even if the occasional patient dies.

Innocent-Enough Beginnings

My own first use of technology for surveillance (or at least for warning purposes) involved a copper-wired alarm I built and placed on my bedroom door. I grew up in a relatively small home, and my bedroom door had the distinct disadvantage of being next to the door of the only bathroom in the house. Inadvertent entries into my bedroom—and, being a younger brother, I have little doubt, occasional *intentional* entries—happened a bit too often for my liking. What's a pre-adolescent boy to do but borrow his engineer Dad's books and put together a D-battery-powered door alarm, replete with a 25-cent copper-wired buzzer? OK, maybe the idea came from my brother's Boy Scout manual—we used *that* book to build an electric motor from scratch, and, DHS-forbid, we used an idea from the same book to generate hydrogen gas using only batteries, copper wire, salt water, and a test tube from my brother's chemistry kit. (Yes, in the process of using a match to ensure the clear stuff in the test tube was indeed hydrogen, we created a minor explosion in Mom's kitchen sink—test tube thoroughly destroyed in the operation. While the test tube was lost, much youthful enthusiasm for technology was gained.)

In any case, there is no doubt that our society, our state, our nation, and our many layers of government provided (or encouraged) a different environment for exploration of technology in the 1950s and 1960s than they provide our young folks today. My test-tube-cracking hydrogen explosion in Mom's kitchen and the delightful true story of the book and movie *October Sky* (including a scene in which author Homer Hickam inadvertently blows up *his* Mom's backyard fence) have little chance of recurring in the Bush-Obama era.[39] Perhaps the new taboos on chemistry sets, model rockets, and such are placing additional distance between our young people and the technologies they will use throughout their lives. It sometimes seems that the more we view technology as "a black box" or as the purview of only a government-sanctioned elite, the more likely we are to encounter confusion (or lack of wisdom) about when, where, and how to use it. It seems possible that when leadership attempts to keep us safe from technology, they may in the long run introduce a different set of dangers based in an incomplete understanding of the tools and systems we use.

But this chapter is about technology-based ambiguation, and we believe ambiguation is a powerful word. If not powerful, at least it represents extreme danger. Ambiguation—the act or process or state of confusion or a lack of clarity—is *not* what we normally want as individuals, as groups, and certainly not as a society or a nation. Ambiguation comes in unlikely forms, and when we ignore it, our ignorance impacts the most unlikely citizens. At the time I built the door alarm for my bedroom, I was an excellent student in school, a strong member of my church and its scouting group, and generally considered to be a decent and "safe" kid. (Of course I know the real score on that point—but appearances can have their advantages.) Despite my reasonable standing within my family and the community, our home maintained a policy of "no locked bedroom doors." It was the 1960s, and I suppose my parents heard a few too many stories about weed-toking kids locked in their bedrooms, puffing away. So I looked for a work-around for the no-locked-bedroom-doors policy, and the low-tech door alarm was my compromise/solution; if I couldn't lock my door, I could at least surprise the heck out of an intruder and alert myself to the invasion of privacy.

This simple example is brim-full of insight into what we do as a society over and over, every single day. We delegate formerly human activity (e.g., communication, action, confrontation, discussion, intervention, negotiation) to some form of technology.

In hindsight, I know that my childhood would have benefited had I taken the time and effort to talk to my parents about my desire for a bit of privacy as I approached my teen years. We could have had discussions about the issues parents inevitably worry about, and I may have had the opportunity to learn how to negotiate on issues of concern rather than deal with them via passive-aggressive use of technology. That sounds like a lot of moral and social depth dug out of a simple story about a boy and his homemade door alarm; however, it does serve to illustrate the simple and beguiling way in which many of us abdicate human roles and action given the opportunity to employ technology instead.

If technology was itself sentient, if the frights (and, dare we say, possibilities) Langdon Winner describes in his classic book *Autonomous Technology* were real, the consideration of ambiguation we present in the following pages could follow a very different path. What if Winner's excellent summary of Mary Shelley's well-known novel, *Frankenstein,* described some near-future reality for the human race?[40] That is to say, what if we could, aside from the God-given abilities for husband and wife to procreate, create sentience, create even a semblance of a soul or a mind in matter that we assemble?

One of our underlying assumptions here—with apologies to Shelley and Winner—is that autonomous technology, insofar as we mean technology that generates decisions and actions without *any* prior form of human intervention, neither exists nor will it ever exist. We will, however, continue to refine our machines (and machinations). We will likely continue to program artificial voices—the best of them already indistinguishable from real voices—to voice more complex and more human-like thoughts, actions, and sentences; perhaps these voices will even one day "create" books.[41] None of this, nor any amount of sophistication of artificially induced locomotion, mimicry, observation, collation, or ratiocination, will amount to humans having created a sentient soul or being. Given that assumption, all human-like activities by machines are, by definition, mimicry. This is a distinct and perhaps headstrong or unimaginative starting point, but a conceptual foundation it is.

To clarify our position, we quote Joseph Weizenbaum, a seminal voice in artificial intelligence:

> Man is not a machine, ... although man most certainly processes information, he does not necessarily process it in the way computers do. Computers and men are not species of the same genus. .... No other organism, and certainly no computer, can be made to confront genuine human problems in human terms. ... However much intelligence computers may attain, now or in the future, theirs must always be an intelligence alien to genuine human problems and concerns.[42]

We immediately encounter the question, "what does all of this have to do with digital reliability?" The answer in this case has to do with both our expectations and our intentions and how those ultimately impact our behaviors. We call on Fred Reed, who discusses not a full-blown

Frankenstein's monster but a much simpler example of ambiguation that begins with mimicking *parts* of human conduct—in this example, self-restraint:

> The scope for automated control of behavior is great. Toyota recently unveiled a car that requires you to insert your driver's license to start it. It then checks your driving record and if, for example, you have a record for speeding, it limits the horsepower that the engine will deliver. (Toyota says it has no plans to put this atrocity into production. Then why build the demonstrator?)
>
> Maybe it's just me, but I'd rather live in a world with less enforcement of laws and more freedom to choose. Years back, this worked. In a society in which reasonable responsibility was culturally mandated, people took laws as guidelines. There were far fewer laws in the first place. The United States is now a country in which personal responsibility is attacked as elitist *and electronic control of behavior seems set to become a substitute.*[43] [Emphasis mine.]

We considered beginning this chapter by recounting a few "war stories" from aviation. We find it much easier to engage attention with stories of self-flying aircraft gone haywire or descriptions of exotic self-guiding ammunition—both of which we know something about;[44] but Reed's down-to-earth observation goes quickly and directly to the core of the issue of technology-induced ambiguation of roles. All of us have roles and responsibilities, and our actions and decisions (and sometimes even more so our *inaction* and *indecision*) impact families, communities, and nations. The ways in which we're currently applying and adopting digital technology into our lives are impacting our entire understanding of what we should do, when and how we should do it, and how (or if) we are even accountable for consequences in given situations.

Referring back to Reed's example of a digitally restrained car, what are the odds that an irresponsible teen driver—the kind very likely to trigger Toyota's horsepower-limiting capability—would make the quite plausible error of believing that he could subsequently drive just about any way that pleased him since the car's monitoring system had already done the dirty work of reducing power to a safer level? After all, drivers who don't think responsibly enough to self-constrain seem likely to make similarly poor judgments about their responsibilities once we've used digital technology to blur the lines of accountability.

What we expect from technology and what we intend to do with it changes so rapidly (i.e., *expands* so rapidly) that our social structures cannot help but be impacted. Joshua Reeves restates Fred Reed's (and our) concerns as follows, as he describes the growing role of the state in first surveilling and then, inevitably, controlling behaviors at both the individual level and for society as a whole:

> Because state-driven lateral surveillance thrives on...ambiguation of citizens' public and private responsibilities...it is a problem that is ripe for historical study alongside the evolving relationships between governmentality and sovereign power that it accompanies. In particular, it is important to examine the potential for this responsibilization to *fracture the social by transforming communal bonds into technologies*....[45] [Emphasis mine.]

"Communal bonds," in our horsepower-limiting Toyota example, could certainly include dads and moms taking car keys away or neighbors shouting out to hotrods to slow down (as I witnessed my neighbor do in the 1960s, resulting in a face-to-face confrontation with the offender). Clearly, responsible parents, and sometimes even friends and neighbors, are in better position than machines to decide whether a few speeding tickets were simply anomalies (or accurate early warnings) that got through to Junior effectively (or not). Digitally enforced power reduction includes no awareness of the human ethical and personal intricacies involved, nor, at a far more practical level, does it guarantee in the least that Junior won't still wrap the family Corolla around a tree.

Reed and Reeves both note the obvious—that we have in our era embraced technology so completely, so whole-heartedly, that even important civic, parental, and moral decisions and activities we abdicate to technology. And, worst of all, the abdication is rarely clear-cut; we move and act in some cases like pliant, quiet citizens, and in others we use the awesome power of surveillance and reporting technology to rein in a neighbor without risking the slightest personal encounter. Reeves describes this Jekyll-Hyde behavior by using the term "citizen/officer/suspect," noting that we freely move back and forth between these roles in our daily lives to suit our intentions (cowardice?) of the moment.

He notes that "we monitor one another not simply because we have the technological resources to do so, but because in the digital age we have developed an assertive, skepticist self-reliance that has eroded our confidence in mainstream social institutions and popular opinion."[46] Perhaps the main point we should derive from Reeves' "eroded confidence" comment is that the impact of ambiguation of roles extends far beyond singular events in which a machine replaces a human in performing a specific task; the larger and far more disconcerting impact may be the collective effect of our constantly growing reliance on machines, which seems to foster a twisted sort of "self-reliance" (to borrow Reeve's wording). This form of self-reliance is marked not by the traditional moral hallmarks of individualism in a free and liberal society (e.g., accepting responsibility for our actions, treating one another with integrity and respect for personal boundaries, voluntarily giving back to our families and communities) but by a tacit distrust and rejection of peaceful communal and familial interactivity. In short, we "do nothing" in the very places and institutions that once provided a very human stability.

And so Reeves is perhaps too generous in labeling our non-responses as "self-reliance." Rather than a positive form of self-reliance, all too often we see resignation ("yes, they can wire up that new intersection camera even while the public library is closing down for lack of funding"), abdication ("I know I should play with my daughter, but the cable's on and we have a baby monitor in the room"), and cowardice/laziness ("it's easier—and less dangerous for me—to snap a photo of his plate with my phone and call 911 than talk to him to find out what's wrong").

Turkle offers similar insight into the real nature of the ambiguation at hand; she notes that growing acceptance of computer psychotherapy—the very concept that frightened her colleague Weizenbaum into decades of resistance to "advances" in his own field of artificial intelligence—do *not* "reflect new abilities of machines to understand people, but [reflect, rather] people's changing ideas about psychotherapy and the workings of their own minds, both [now] seen in more mechanistic terms."[47] In short, we see the harmful effects of ambiguation of roles in action, as computer psychotherapy reflects resignation and acceptance of technology in place of humans, but this does not represent anything genuinely new and improved in terms of our ability to understand and mitigate our own harmful attitudes and behavior. This particular form of ambiguation may deeply exacerbate the very types of problems we attempted to solve in the first place; only extreme behaviorists are likely to believe that viewing ourselves "in more mechanistic terms" will improve human psychology.

And perhaps Turkle here identifies what should be our key concern about ambiguation: the machines we invent cannot "worry" about their potential intrusions and trespasses into human affairs. They certainly have no ability (nor should we ever hope that they do) to police themselves at a conceptual level, to measure their own impact on things social, psychological, and communal. We—their inventors—have that inherent responsibility and should once again embrace it. To the degree we discover ambiguation of human roles upon the introduction of a new technology, we have the obligation to try to *continually recognize* that ambiguation and to work to understand, counter, and eliminate its negative impacts. And yet we often act as if new technology is a Pandora's box—unstoppable upon introduction to the world. This attitude, in concert with our propensity to embrace determinism and our willingness to reify history (or progress or any number of other human concepts), contributes to larger failures to stop new technologies at their inception. Transcripts of attempts to collectively outlaw building nuclear weapons in the 1940s show graphically that even with worldwide desire to stop a given technology, we usually fail to restrain or halt technologies, even the ones we fear most.[48]

Sadly, in my own life and attitudes and in the language I use, I adopt, often unwittingly, many of the foibles described above, including thinking deterministically and reifying concepts that cannot possibly demonstrate independent sentience. So our intent here is not to supply a hypocritical sermon but to help identify how digital technology fails us—where it is unreliable within the parameters of our expanded definition of digital reliability—and also to remember that "rather than being 'outside' society, technology is an inextricable part of it."[49] Since both society and technology exist (and continue to exist) through human volition, we should more accurately describe our effort here as being to help *identify how we fail to notice, monitor, and moderate the impacts of technology*. How and where do we fail to engage our neighbors, leaders, our managers, and our institutions by asking questions such as "have you accurately and completely assessed this use of technology before committing my taxes toward it?" Or even our pastors: "Have you considered whether or not filming that sacrament for replay on YouTube is consistent with the meaning and intent of the sacrament itself?"

Simplistic? At this point, it is most certainly simplistic and naïve to believe we can expect ourselves and our neighbors to walk into a town hall meeting (they do still have those?) and ask the local leadership whether or not the new intersection in town really needs a license-plate reading camera. People today don't ask those types of questions, or in the rare cases they do, they do so knowing their names will likely, in short order, make a place on a politically and/or financially emasculating list somewhere soon after the town hall meeting ends. (If you don't think that sort of thing happens, read Greg Richter's article that reveals the horrific political manipulation of the IRS under the Obama administration also extended to other agencies and institutions that administer and oversee our finances and communications.)[50]

Even so, and notwithstanding all the aforementioned warnings and cautions, this ambiguation, and our ambiguous reactions to it, are, at times, not without positive motivation and not without moments that help us to claim the moral high ground. Reeves, for example, describes the actions of a neighborhood watch program that played a part in cleaning up an illicit-drug-laden Los Angeles neighborhood; the local citizens offered the police an impressive list of drug-dealer behaviors they had recorded themselves, including times and locations when dealers traveled over local fences and what they were wearing that day.[51] These citizens acted, with the deepest of motivation, as "officers" in an effort to reclaim their own community. They clearly believed the police could not accomplish their community goals without the work of technology-aided citizens providing real-time information. This points to the inherent complexity of issues involved when society receives wave after wave of devices and technologies that, Frankenstein-like, confound human responsibility (and action) with artificial and powerful machine-aided actions and activities. For every 100 times we seem confused, disappointed, de-motivated, or victimized by those wielding a new technology, we can find the occasional example of "technology to the rescue."

We could, of course, offer far less noble examples of citizen-officer-suspect ambiguation—even examples from eras far before the advent of digital technology. Reeves notes that in the Norman system (a thousand years ago), since "individuals were made to remain watchful of everyone in the village, not only did the threats of crime and its response become ubiquitous, but the community's sociality was now forced to revolve around disciplined rituals of mutual suspicion."[52] These "rituals of mutual suspicion," Bentham's and Machiavelli's ideas aside, do not, in our opinion, provide a foundation for any society other than one resembling hell itself. Indeed, Reeves, finding deep parallels between the Norman system and our own, calls our current environment a "climate of categorical suspicion":

> DHS and other security agencies are increasingly vocal in their efforts to encourage vulnerable citizens to become vigilant in the fight against terror. And because all citizens in the post-9/11 era have been declared potential walking/driving/flying bombs, this vigilance against terror translates into vigilance against one another.[53] The secrecy and potential ubiquity of terrorists, argues Jeremy Packer, 'creates a situation in which combatants cannot be known in any field of battle, which means everyone will be policed as if they are potential terrorists. At the same time, all citizens are asked to join in the war on terror as part of DHS initiatives.'[54] The ubiquity of this threat is a key impetus

of public lateral surveillance, further ensuring the ambiguation of individuals' citizen/officer/suspect roles. When everyone has become a potential suspect, there is no way that the police can keep up with its surveillance demands. But if lay citizens can be encouraged to watch one another—not only when using Facebook or Twitter, but when they shop at a supermarket or peer out their kitchen windows—the official representatives of the governmentalized state, like the constables in Anglo-Norman England, can be freed to devote themselves to activities that have little to do with community improvement.[55]

Other Forms of Ambiguation and Real-life Examples

The social, psychological, and political ramifications of technology-aided ambiguation of roles just discussed are compelling, but we should also mention some of the more commonplace (or easy-to-recognize) impacts of ambiguation. We've all heard of smoke alarms that woke would-be victims and saved homes and lives. Sadly, many of us have also heard of children lost to home tragedies because parents over-relied on tech-nannies of various kinds (monitors, alarms, or simply an ever-running television).[56]
I encountered the issue of over-reliance on technology (a form of ambiguation of roles) in aviation more than once, but my first encounter was perhaps the most memorable. We were approaching a city with a load of passengers in back and a bank of very gray and heavy clouds in front—thick clouds right between us and our destination airport. I had flown into this city many times before, and I knew that part way down the approach to the airport, the tallest smoke stack I'd ever seen stood waiting for some pilot in bad weather to forget to watch the plane's altitude. In short, if you were in the clouds (and, boy, on that particular day, were we ever), you had better make sure to keep above a certain altitude until you passed the chimney. I'd flown this airport's approach in plenty of good weather too, and every time I passed the chimney on a sunny day I'd shudder thinking how dangerous that huge smoke stack would be for pilots flying in "instrument conditions" (i.e., flying with poor or no visibility).

On this particular day, that was us. The view out the front of the aircraft looked like the proverbial "inside of a ping pong ball." But our crew knew the airport, knew its approaches, and certainly knew how alert we needed to be while flying in instrument conditions (i.e., with no visual references outside the plane to guide us). We had the aircraft trimmed and stable, flying at the right airspeed and altitude—practically flying itself, as the digital technology in a modern aircraft enables them to do. Then a small warning light came on. The light had little to do with our direct safety—it did not indicate any problems with engines, aircraft stability, landing gear, fire or any other imminently threatening issue. But the light did indicate something was amiss with one of our navigational instruments—one of several instruments we'd use to help us know when it was safe to descend through the thick clouds.

My copilot knew how badly we needed to ensure we started our descent *after* passing the towering chimney. He began working on the malfunctioning navigation instrument. Reset after reset. Still the pesky warning light would not go away. More crew members got involved; after all, the aircraft was stable, trimmed, and holding a steady altitude and airspeed, so why not address the problem at hand? In short order, I noticed the whole crew, me included, was looking down at this glowing warning light on a backup navigational system; and in fact some of the passengers were even getting in on the discussion. That's when a bit of some excellent safety training came to the rescue. I turned off the intercom (ending the unhelpful discussion with the passengers) and reminded the copilot to "fly the plane first"–i.e., I reminded the whole crew to keep our noses (and eyes, ears, and topics of conversation) out front, where we were headed. I reminded them we had backup systems that could tell us when it was safe to descend, and that we would write up the malfunctioning navigation system for maintenance when we got on the ground. In short, I brought to an abrupt end *the ambiguation* that our stable and autopilot-assisted aircraft was enabling.

That story, which shows how easily ambiguation of roles in modern aviation becomes a factor, ended well—we landed safely and lived to tell the story another day. When an aircraft can practically fly itself, as ours could, aircrews can easily forget their own first priority (as we momentarily did), which is to fly the plane. Troubleshooting, communicating, repairing—even navigating—all have to take second place to keeping the aircraft in a safe, flyable configuration at all times. Sadly, another aviation story, the lessons of which helped me through that situation near the cloud-enshrouded smoke stack, has a much more somber ending.

Imagine looking out the window as a passenger of a brand new wide-body jet that just took off from Miami. It's a dark night, but even so, when the everglades get within a few dozen feet of the underbelly of the plane, you notice the water and trees coming up—far too close, as the plane is way too low, and there's no runway in sight. Then the aircraft starts to come apart around you as a perfectly good, perfectly flyable plane noses into the water, almost as if coming in for a landing in the middle of the everglades—on autopilot.

Some passengers survived that horrific accident in 1972, but many died right there in the swamp. Once the National Transportation Safety Board (NTSB) unwound the details, the tragedy seemed all that much worse; the flight crew and the aircraft (including its autopilot system) were in just about perfect condition when the plane hit the water. Had any one of the crew noticed the proximity of the water even a few seconds before impact, the plane might still be in flying inventory today.

What happened that night was that the crew noticed a burnt-out $12 light bulb on their landing gear position indicator and decided to collectively turn their attention to that bulb and work on replacing it—while the autopilot held their altitude and airspeed stable. This was a classic case of technology-induced ambiguation of roles, since piloting the plane is always, without fail, the number one priority in aviation.

Somewhere in the middle of the crew's huddle to fix the bulb, one of the crew bumped the jet's steering yoke forward. The jetliner's autopilot did exactly what it was programmed to do in that situation—it interpreted the "bump" forward on the yoke (which would nose the aircraft downward) as a command to start a controlled descent. The electronic controls in the autopilot completed their task in that case exactly as designed (i.e., they executed with *task reliability*), holding the slow angle of descent and the aircraft's airspeed stable, all the way into the swamp. The crew, confident the plane's technology systems had everything under control, never noticed the descent, never sounded an alarm, never looked up from their focus on the burnt-out bulb, and never even had the chance to pull the yoke back to stop the descent.[57]

When this tragic crash story is recounted during flight safety lessons, instructors often focus on the "lost jumbo jet for want of a replacement light bulb." It makes a catchy storyline, for sure, and that day as I piloted my plane in thick clouds near the smoke stack, I can honestly say this 1972 crash story came to my mind and helped me refocus myself and the crew on "flying the plane first." We miss the larger point of the crash in Miami, however, if we focus on identifying the crew's errant fixation on the bulb as the main irony here. Even focusing on the autopilot and its possible shortcomings is not getting to the root of the problem, although that airliner accident fortunately led to the addition of altitude-warning systems designed to help avoid a similar incident (thesis, antithesis, synthesis again—as we'll discuss in Chapter 6).

We maintain, though, that there is more to solving and preventing tragedies like these than improving technology. Due to the complexity of jet aircraft, engineers provide aircrews with many sophisticated systems that mimic the actions of human pilots. Most of us are familiar with an automobile's "cruise control" system; modern aircraft provide pilots with that type of assistance in spades. Most high-end autopilots maintain heading, altitude and airspeed far better than a typical pilot can. What happens, then, when pilots, whose first mandate is always to maintain their aircraft in safe, controlled flight, begin delegating—giving away—their physical and mental activities to pilot-assisting systems, such as autopilot? In the NTSB's own words, here is a sequence of behaviors and events that can unfold when we begin delegating piloting tasks to digital technology:

> [NTSB] investigators believe the autopilot switched modes when the captain accidentally leaned against the yoke while turning to speak [about the burnt-out bulb] to the flight engineer, who was sitting behind and to the right of him. The slight forward pressure on the stick would have caused the aircraft to enter a slow descent, maintained by the CWS system.[58]

"Provide redundant or 'infallible' warning lights"; "Improve autopilots!"— these are the types of suggestions and demands that come out of an accident like the one near Miami; they also fit neatly into the category of solutions Langdon Winner alludes to when he says that we "deal with the specter of vulnerability" by trying to "ensure that technical devices and systems are well-engineered and protected from calamitous failure."[59] Winner, Talbott, Neil Postman, and others, however, have noted that the more vital issue, and an issue that gets very little press and even less funding, has to do with the fact that as we build more and larger complex technical systems, we delegate our former roles (and attention, and focus, and even our skill sets) elsewhere—somewhere other than the task(s) at hand. At certain points in this process of divesting responsibility, Winner notes, we encounter a tragedy that gives us pause. Our reactions in those cases, however, are far too often to simply redouble our efforts to create safer, ever-more integrated systems—and nothing more or significantly different. Our attitude is too often that "technology created this problem, and technology will fix it."

Again, though, we should ask ourselves if task-performance issues should remain our focal point in analyzing technology-related failures; sometimes the "tragedy" we encounter in ambiguation of roles is far subtler and more difficult to discern than airliner crashes, and sometimes the tragedy unfolds slowly over years or even decades. The "reliability problems" we try to solve typically involve determining if technology is completing accurately a set of tasks, such as holding a plane's altitude and airspeed, calculating pi, and monitoring our homes and driveways for intrusion. But if we agree that the real question at hand should be "How is our use of technology actually affecting us?" then the urge to "fix" a situation by improving digital reliability for a given set of tasks may be completely off the mark and may even exacerbate the original problem.

For example, recent South Korean research into the effects of pervasive digital tool use by young people led to the coining of the frightening term "digital dementia." The very tools people adopted to help make them more aware, informed, and capable of making "smart" decisions led to a frightening crippling of those cognitive abilities in the users of those tools:

> Doctors have reported a surge in cases of digital dementia among young people. They say that teenagers have become so reliant on digital technology they are no longer able to remember everyday details such as their phone numbers. South Korean experts have found that those who rely more on technology *suffer a deterioration in cognitive abilities more commonly seen in patients who have suffered a head injury or psychiatric illness.* ...'Over-use of smartphones and game devices hampers the balanced development of the brain,' Byun Gi-won, a doctor at the Balance Brain Centre in Seoul, told the JoongAng Daily newspaper. ... Sufferers are also reported to suffer from emotional underdevelopment, with children more at risk than adults because their brains are still growing.[60] [Emphasis mine.]

Ambiguation of roles is clearly more than just a problem for airline pilots. Stephen Talbott describes our actions and processes when ambiguation of roles is at hand. He notes a point that seems at first blush quite comforting—*if* the only type of problem we're considering is preventing a jumbo jet from plowing into a swamp while its computer systems silently execute their assigned tasks to perfection: "No law seems more certain than this one: the next generation of computers will be better than the last." Who wouldn't want aviation computers to be more "aware" of context and danger than the ones that went quietly and steadily into the water with hundreds of passengers that December night? Talbott then drops the other shoe: "Yet no law conceals a more socially devastating lie."[61] As we accept more and more digital technology into our lives, as we integrate our lives with machines, the temptation to confuse responsibility (i.e., the temptation to ambiguation) grows. Talbott advises:

> Do not blame the computer. If you and I are satisfied with products and services that bear no moral, artistic, or purposeful imprint of

another human being, then the computer is the perfect instrument of delivery. ... But keep in mind that this 'progress' reflects not so much the machine's increasing friendliness and humanity as it does your and my willingness to become more machinelike.[62]

What do "sociality," "friendliness," and "humanity" have to do with airliner crashes (and, ultimately, digital reliability)? As it turns out, more than meets the eye. Flight safety, for instance, has enjoyed a tremendous boost in recent years thanks to the use of aircrew coordination training. US Army aviation became interested in aircrew coordination training after they noticed that accidents were occurring over minor aircraft malfunctions (such as, in the example above, a burnt-out light bulb). They found that while their "aviators displayed excellent understanding of aircraft systems, operating procedures, rules and regulations and other technical information, they often displayed a glaring inability to communicate effectively, distribute workload, maintain or regain **situational awareness** and make sound decisions."[63]

Maintaining **situational awareness**—perhaps the single most important key to safe and effective aviation—is both helped and hurt by aviation technology. Before the invention of radio-assisted navigational aids, instrument-based flight through clouds, rain, and snow and into unfamiliar cities and airports, day or night, was either impossible or impossibly dangerous. Technology, particularly navigational and communications technology, indeed aids the pilot in maintaining situational awareness in many ways. Just as certainly, some uses of technology in aviation discourage human communication and action, divert attention, and lull pilots into complacence just when vigilance-based **situational awareness** is most in demand.

Aircrew coordination training, with its emphasis on the human communication elements of aviation, made an easily traceable impact in flight safety over the last 25 years. True, aircraft manufacturers also improved navigation systems and flight computers over that period, but evidence shows that simply improving the human communication aspect of piloting aircraft contributed mightily to increased safety. In a sense, this is a rare example of our actions—in this case, refocusing aircrew training—reducing or reversing the negative aspects of technology-induced ambiguation of roles. The US Army, and later, most commercial airlines, began to recognize that ever-improved digital technology would not take the place of human pilots who were trained to better understand their real roles in the cockpit and to recognize the pitfalls and inappropriate crutches that we sometimes introduce inadvertently with high-end technological aviation assistance.

I once had a discussion with an instructor pilot about another aspect of "humanity" in aviation. A military academy graduate, this instructor was of the opinion that current military-uses of aviation were impacting the ancient warrior/poet model of the soldier. I did not see his point of view initially, knowing next to nothing at the time about the warrior/poet model. I was of the opinion during my days in the cockpit that if we needed to stop the enemy in defense of our land and kin, then having the heart of a warrior/poet was far less important than, for example, having an accurate bomb sight and good maps.

He tried to explain to me the difference between engaging one's enemy face-to-face as opposed to hurling explosives at him from a distance; he quoted others who thought the modern bomber pilot to be one of the greatest cowards of the battlefield. (We had this conversation long before person-specific drone strikes conducted from unseen heights and controlled from a great distance via digital technology were all the rage.)

We won't tackle ambiguation of roles in the case of drone strikes here, in part because the idea and practice are still very new. We make note, though, that we should probably collectively ask more questions about this type of use of technology than just about any other technology we can think of. Drone strikes give the power of life and death with no physical threat to the attacker, absolutely no danger to the controller(s) and decision makers, and, for now, a minimum of even political accountability. This technology cuts across an incredibly wide spectrum of human moral and spiritual functionality, and it does so with the appearance of complete immunity from repercussions—and that is to say, repercussions of *any* type; we believe the potential for technology-induced ambiguation when using drones to patrol and strike human beings is far greater than even that associated with nuclear missiles. That is not hyperbole—just an opinion backed by many years of experience in both military aviation and digital technology.

In summary, technology-induced ambiguation of roles is an intimate, powerful, ubiquitous, and yet hard-to-notice feature of our epoch. It impacts aviation, child care, the policing of our streets, and virtually all aspects of our lives because we have welcomed technology *into* all aspects of our lives, and we have done so with surprisingly little examination of potential repercussions.

5 CONTROL CENTRISM

Our time prides itself on having finally achieved the freedom from censorship for which libertarians in all ages have struggled...The credit for these great achievements is claimed by the new spirit of rationalism, a rationalism that, it is argued, has finally been able to tear from man's eyes the shrouds imposed by mystical thought, religion, and such powerful illusions as freedom and dignity. Science has given us this great victory over ignorance. But, on closer examination, this victory too can be seen as an Orwellian triumph of an even higher ignorance: what we have gained is a new conformism, which permits us to say anything that can be said in the functional languages of instrumental reason, but forbids us to allude to...the living truth...so we may discuss the very manufacture of life and its 'objective' manipulations, but we may not mention God, grace, or morality.

— Joseph Weizenbaum, Computer Power and Human Reason: From Judgment to Calculation

Neil Postman, in his book Technopoly, notes that "computers are far more 'human' than...almost any other kind of technology. Unlike most machines, computers do no work; they direct work. They are, as Norbert Wiener said, the technology of 'command and control' and have little value without something to control."[64]

In some ways, control-centrism and its eruption via our broad and deep embracing of the technology of command and control is the "uber issue" in our discussion. Much of our analysis about misguided thinking and dangerous paths in the digital realm starts with analyzing digital technology's foundation in the age-old problem of people wanting to control what others are doing and, increasingly in our epoch, to control what others are thinking.[65] Wanting to control others by any means, digital or otherwise, is an ancient problem, perhaps at the core of the first recorded murder in the Bible. The prayer of St. Ephraim, a staple of liturgies celebrated by millions of Orthodox Christians around the world, even identifies "lust of power" as one of the four basic sins of mankind.[66]

If the desire to control others is indeed an ancient problem, what then, if anything, makes control-centrism in an era of digital technology different? Are we, in our time, dealing with a quantitative difference, as in "more" desire to control others? Or should we look for a qualitative difference, as in the ability to control others is somehow different in a digital era than the unmistakably powerful social control exerted by, for example, the ancient Romans? Even a brief overview of the ability of Roman rulers and legions to control their masses impresses. The array of tortures and "social" events (if public displays of gruesome deaths by torture can be classified as social events) is rivaled only by their layered legal and military systems in place to accuse, catch, and sentence offenders. So—did pundits, critics, and writers of the Roman era see their new technologies (road building, improved transportation, advances in metallurgy) as a core cause of social control extremes in Roman society? Is there evidence of a standing correlation between quick advances in technology and the tendency of a given era or empire to trend in the direction of social control?

Our own era in the US at least rivals ancient Rome quantitatively for its focus on control of the masses. Pew Charitable Trusts produced a study showing that more than 1 in 100 adults in the US are incarcerated at any given time—more than any other nation on earth, and perhaps the highest incarceration rate ever recorded.[67] Incarceration, however, is only one measurable aspect of our control-centric behaviors and attitudes, and at the moment, the impact of digital technology is far more evident in incarceration's unlovely twin, surveillance.

But, but...Things Are Different Today!

Before we take a brief look at the current state of surveillance, we'd like to acknowledge one very strong—and repeatedly encountered—argument in favor of our current penchant for maintaining a level of citizen surveillance far above the usual norms for a state, nation, or administration. Many argue that technology as we know it today gives the individual citizen an amplified ability to inflict harm, both to his fellows and to the state as a whole. While we do not deny that a cellphone-triggered improvised explosive device (IED) possibly brings nearly as much newly frightful destructive capability as the invention of gunpowder did in the 9th century AD, modern states *do* routinely look the other way when powerfully dangerous new technologies emerge.

They certainly look the other way when the general interest of "progress" or "civilization" appears to be at stake. John Taylor Gatto explores this point beautifully and in depth in his classic book *The Underground History of American Education*. Among other deeply telling points Gatto makes is that all nations currently permit the use of gasoline-powered vehicles, meaning that thousands (and in many larger countries, *millions*) of citizens routinely drive vehicles weighing thousands of pounds loaded with many gallons of highly explosive liquid, and they do so every day. And they have done so for over 100 years.

Gatto also notes states that otherwise display high regard for the general safety of their citizens routinely turn a blind eye to the deaths of hundreds (or even thousands) of men each year in the coalmining industry.[68] States that, in the name of security, will not even permit completely free use of the Internet knowingly lose 1,000 or more men in coalmines each year. China lost almost 5,000 men in the coalmining industry in the year 2003 alone.

As Gatto describes so poignantly, states, and the wealthy elite so often standing at the top of their political structures, have at best inconsistent motivations and safeguards in place for monitoring and controlling dangerous technologies (and the means, techniques, and tools for the acquisition of those technologies). We won't recap all of Gatto's arguments here, but will recommend heartily his work and insights on this point.[69]

State Surveillance and Its Modern Roots

For the sake of taking a penetrating look at the status of state surveillance, we'll temporarily set aside the disclaimers and justifications of those who would monitor all in the name of national security, or security in general. Jeremy Bentham and his Panopticon provide an appropriate starting point for discussing some of the historical foundations and subsequent implementations of surveillance in the modern state.

Bentham developed a prison design (the Panopticon) meant to enable guards to physically observe inmates at all times. Many refer to Bentham's ideas and works in discussions about our current stance toward surveillance. Indeed, in the UK and the US, governments seem to have created a default acceptance of the state Bentham described as his ideal, which was derived from the Panopticon:

> Bentham argued that those inside the panopticon should always think they are under inspection at any time. It is important 'that the persons to be inspected should always feel themselves as if under inspection, at least as standing a great chance of being so'. Not only are the prisoners under the possibility of constant surveillance but the panopticon also serves to keep 'the under keepers or inspectors, the servants, and subordinates of every kind, to be under the same irresistible control with respect to the head keeper or inspector, as the prisoners or other persons to be governed are with respect to them.' [70]

As Tom Brignall III notes, "Bentham believed in the omnipotent potential to be observed at any time." Bentham wrote his ideas, of course, well before the advent of digital technology and long before leaders of the NSA

61

decided that collecting data on the phone calls and communication of every single citizen would improve the world.[71] Even so, his ideas affected much of the world's social fabric, influencing the designs of prisons and many other places where citizens might gather and potentially providing effective ideas for despots and would-be dictators everywhere.[72]

Bentham's design relied upon visual surveillance (i.e., he proposed the use of low-end, non-digital technology), and although his Panopticon centered on observation of the criminally convicted, as Brignall III notes, Bentham meant for his ideas to influence the design of virtually all physical buildings and places in which large groups of citizens congregate. His motivation: surveillance as a means of social control. He said that his design provided "a new mode of obtaining power of mind over mind, in a quantity hitherto without example."[73] "Mind over mind"—note the use of a physical design (or infrastructure) with intent to influence *mind*. (Was Bentham a forerunner of the concept of the mechanized mind?)

The design of Bentham's proposed prison building placed a guard in position to view prisoners in lighted cells and living spaces from the vantage point of a darkened tower. This ability to observe without being seen multiplied the effectiveness of the guards' surveillance. I remember driving in a former Warsaw Pact country in the 1990s, shortly after the people of Germany had torn down the Berlin Wall. I noticed that on many key highways, the planners had designed the intersections in a "T" shape, with a tall police tower, similar in design and in its use of glass to an airport control tower, placed right at the top of the "T." This deliberate design ensured police could view all oncoming drivers for a great distance in any direction. The glass, of course, was mirrored, meaning drivers never knew when they were being observed. One day, I drove for miles on a perfectly straight road headed directly at the top of the "T" intersection. As I drew near the very visible "control tower," I remember looking up at the mirrored, opaque windows and thinking, "Bentham couldn't be happier."

As Bentham's ideas flourished, engineers began combining numerous variations of digital technology with video, audio, and even olfactory sensing to provide those interested in surveillance a far more powerful opportunity to observe the activities of others. For example, in the mid-1990s, even as the US was redefining many of its former enemies as new friends, the US DoD requested proposals for audio and video sensing devices that could be placed in remote locations as faux rocks, trees, ice, and other naturally occurring objects for the purpose of surveilling those who believed they were free from observation. This means the DoD (and other agencies) could surreptitiously monitor conversations and activities in the desert, the woods, or the arctic. Clearly, even peace does not dampen the motivations of some state leaders to surveil at all times and in all possible places.

That said, few dispute that state-sponsored surveillance, at least in some forms and some instances, can and does have a place. Perusing documents from the Cold War shows that leaders have long promoted forms of surveillance in the name of openness and security. A 1991 paper from the Office of Technology of Assessment dealing with issues related to the fall of the Soviet Union called for renewed emphasis on a mutual surveillance concept initiated by President Eisenhower as far back as 1955. In discussing Eisenhower's newly revived Open Skies proposal, which called for *mutual and consensual* surveillance of countries, the report has chapters with innocuous names like "Open Skies?," "Airborne Platforms and Sensors," and "Why Aerial Surveillance." While we can hardly imagine a Soviet or Russian citizen looking up at a US surveillance aircraft packed with the latest digital surveillance technology and thinking, "Geez, I feel better," the report mentions surveillance in nearly glowing terms, talking about "mutual confidence," cooperation, and good faith. At one point, the 1991 report states:

> The Open Skies Treaty, which is now being negotiated by members of NATO and the now-disbanded Warsaw Pact, is intended to be primarily confidence-building measure [sic] to reduce international tensions and foster trust and goodwill. ... Americans, in concert with others, may someday be able to fly aircraft through the Soviet Union and other countries on a reciprocal basis, taking pictures and collecting other data that will contribute to a more secure future.[74]

Again, while we can occasionally justify state-sponsored surveillance even under a Biblical umbrella (e.g., the Israelites sending spies into the land ahead of their migration to a new homeland), the thought that serious thinkers and leaders, such as those who contributed to the report quoted above, actually claim to believe that *mutual and consensual monitoring of sovereign states is an activity favorable to a sense of trust* clearly indicates how deeply embedded our current surveillance state mindset is. There was a time, not very long ago, when citizens would have thought writers of such ideas had less-than-balanced minds.

Angela Merkel of Germany and Dilma Rousseff of Brazil give us a glimpse of what political elites really think about a modern state taking away privacy via technology; consider their reactions upon finding out the NSA was having its way with *their* cellphones and email.[75] They certainly did not hail the revelation as a new form of international openness, nor did they suggest we all embrace an "Open Internet Treaty," based upon mutual and consensual email monitoring in the spirit of the "Open Skies Treaty." Hillary Clinton—serving as Secretary of State in an administration that made active use of purloined email communications to make state decisions—defended her right to be "the arbiter of which e-mails to keep for archival purposes and which to delete" when confronted with her own private email crisis in 2015. "No one wants their personal e-mails made public, and I think most people understand that and respect that privacy," quipped Clinton.[76] We couldn't agree more on the need to understand and respect, and we wonder (tongue in cheek) whether Clinton, consequent to her own experience with invaded privacy, now champions Snowden's whistleblower cause?

In summary, combining a mindset that the state must observe as much human activity and communication as possible with digital technology's incredible augmentation of the ability to do so brings us to the next trend that impacts our ability to correctly judge, measure, and understand digital reliability: The Participatory Panopticon.

The Participatory Panopticon

"To be blunt: the more that people feel like these tools are extensions of themselves, the less they'll want to have them restricted," says Jamais Cascio about devices designed to facilitate and simplify digital video recording and distribution in his essay, "The Rise of the Participatory Panopticon." "The proliferation of small, easily concealed and readily networked digital cameras can be a headache for those trying to retain some degree of privacy [e.g., Merkel, Rouseff, and the rest of us], but they're a nightmare for those trying to keep hold of some degree of secrecy [e.g., LAPD, NSA, and the rest of them]."[77] Remembering that Bentham himself wanted, in addition to prisoners of the Panopticon, "keepers or inspectors, the servants, and subordinates of every kind, to be under the same irresistible control [as the prisoners of the Panopticon]," the idea of a society-wide Panopticon in which we all willingly play a part, mutually observing and reporting on one another, seems a natural progression of Bentham's odious ideas.

Again, what does Cascio's concept of the "Participatory Panopticon" have to do with digital reliability? It's a good question, with we believe a great, albeit less-than-obvious, answer: Measuring performance on specific tasks will always be a *part* of how we quantify digital reliability, but we state again that measuring digital reliability accurately often means more than gauging how well hardware and software perform *only* the specific tasks they were designed to do.

When considering the *specific individual tasks* we assign to technology, sometimes noticing a deficiency of digital reliability is simple and straightforward. This is perhaps especially so in a field in which we must act with much precision and dependability, such as the field of nuclear technology. When Tokyo Electric Power Company (TEPCO) notes that a rat near a switchboard may have caused them to lose nearly a full day of critical cooling capability at their crippled Fukushima nuclear plant, we can easily think of relevant questions about the tasks, equipment, and systems, as well as the specific situation involved—i.e., did TEPCO centralize so much critical functionality via digital technology that they ended up, quite ironically, increasing their vulnerability to loss of control?[78]

Yet issues and events involving digital reliability are rarely so straightforward to diagnose, and even more rarely are the full impacts of digital technology evident in a single success (or failure) of digital technology. Evgeny Morozov noted during his 2009 TED talk that confusing "the intended versus the actual uses of technology" even leads to the extreme confusion of believing that we can watch democracy and political freedom blossom simply by "dropping iPads not bombs."[79]

But how does a society reach such ridiculous gaps in sound thinking seemingly overnight? We'd like to blame politicians who masquerade control-centric surveillance in the name of "open skies" and "mutual cooperation." But even that legerdemain is not enough to engender the whimsical belief that enabling communication via tech tools will automatically and simultaneously lead to improvements in political systems.

When we consider, however, Morozov's caution about "intended versus actual" uses of digital technology, and when we consider how many millions of citizens use digital devices daily and without reserve, we see a much more powerful opportunity for confusion. Or, more specifically, potential for overlooking unintended impacts and the consequences of our use of digital tools and toys.

That said, here is what we believe to be a vital adjustment to assessing and defining digital reliability. Promoting digital reliability, at all times and in every situation, should also entail:

- Considering whether a given digital technology, taken as a whole, is hurting ourselves or others.

- Measuring and documenting where digital technology is failing us in previously unnoticed, unobserved ways.
- Determining where digital technology is inducing indifference, inactivity, or incapacity where human action should be.

If our new definition of digital reliability seems to represent an abundance of caution, we refer back to the quotation with which we opened Chapter 3: *"Rather than being 'outside' society, technology is an inextricable part of it."* Langdon Winner carries this thought further:

The beguiling but ultimately mistaken notion that technologies are 'merely tools'—things we pick up, use and then easily put away— poses a major barrier for understanding how we live today. Missing in the tool/use perspective is acknowledgment of a basic fact about people's relationship to the technological realm: our utter dependence upon the large, complex, artificial systems that surround us on every side, giving structure to everything we do.[80]

Winner helps us see the Catch-22 in which we find ourselves. We typically accept without comment the proclamations of tech cheerleaders who proclaim endlessly that "technology xyz" will change how we live, how we think, how we communicate in our most intimate settings, etc. Then should we not suggest that anything with such power and potential to disrupt our lives and norms should certainly deserve our deepest attention and caution?

Anything we consider adopting that has the potential to impact our inner lives or our communal existence deserves close examination. The author of a recent study about unexamined uses of meditation said this about our careless (and/or nonexistent) research in *that* field: "The shortage of rigorous statistical studies into the negative effects of meditation was a 'scandal,' Dr. Farias told The Times. ... 'This shows a rather narrow-minded view. How can a technique that allows you to look within and change your perception or reality of yourself be without potential adverse effects?'"

Digital technology may not, in most cases, affect our inward selves as directly as does meditation, but taken as a whole, our outright surrender to "digital progress" might have even more impact on our inner lives and our communities than does meditation. Dr. Farias' outrage about the lack of thoroughness in his field of study, without any doubt, should apply to how we handle digital technology as well.[81]

Winner gives us a glimpse into how we fail to focus on the more genuine impacts of digital technology. When we give *any* thought at all to digital reliability, we typically focus, predictably enough, on good engineering and those headline-grabbing "calamitous failures:" Winner notes, "There are several ways that our society routinely deals with the specter of vulnerability. One strategy is to ensure that technical devices and systems are well engineered and protected from calamitous failure" [e.g., ensuring rats can't take out a nuclear facility].

He then adds:

> But good engineering is only part of the story. In free, democratic societies there is another way in which ordinary people have managed their relationship to vulnerability: they embrace an attitude of trust, holding on to the reasonable expectation that key technologies will always work reliably and not break down *in ways that jeopardize our health, safety and comfort.*[82] [Emphasis mine.]

We embrace Winner's expanded view of reliability, and we hope to reinforce his position by reiterating that the trust in technology we adopt in our society (and are broadly expected to embrace continually) applies to far more than the question of whether technology "breaks down" or "works reliably."[83] When we grant genuine trust, we usually consider ends at least as much as we observe means or technique; trust, as we are asked to render it in a society laden with today's tools of surveillance and control, requires a much deeper consideration of ends, means, technique, and, without a whiff of hyperbole, consideration of a myriad of potential side effects and unforeseen issues and consequences. We are, after all, talking about machines when we discuss digital technology. And today, with machines sifting through our most intimate communications and grading our students' SAT essays, *the concept of machines judging human action and behaviors is included by default.* That thought alone should frighten us into action.

Yet, strangely, our current national emphasis is not on monitoring and mitigating the negative actions and (inevitable) failures of our machines and technologies; far from it—our current emphasis seems to be on increasing our use of machines and technology, even deeply flawed iterations of these, to monitor and control one another.

As we mentioned in Chapter 2, we now embrace robotic surgeons, which conduct the most human-invasive procedures, short of mind reading, and we do so "by and large without the proper evaluation."[84] These are not by any means the most personally invasive use of machines we tolerate collectively; we also enable the most massive and globally active invasion of privacy in the history of the world on the behalf of the NSA and CIA, reined in (if at all) by only a tiny rubber-stamp court whose decisions to permit these invasions (*in a purported republic*) are kept intentionally hidden from the taxpaying public that funds the very existence of the NSA. According to the sworn declaration of NSA whistleblower (and former NSA employee) William Binney, after the events of 11 September 2001, that court, known as FISA, "...ceased to be an operative concern, and the individual liberties preserved in the U.S. Constitution were no longer a consideration [in the NSA regulatory process]."[85]

Interestingly, as Cascio and several other tech writers explore the participatory Panopticon, they make clear that they believe aspects of the concept are quite positive. Tim Rayner, for example, clearly aware of Bentham's late-blooming but ongoing influence, penned an essay exploring how we are adopting social media into every aspect of our lives; nothing unusual about that observation, except for the fact that he also casually explores Bentham's Panopticon as a known facet of our new and blossoming use of social media, even including a picture of a Panopticon-like prison in his article. He does not even play down the aspects of the Panopticon's influence that are almost universally panned as negative, such as Bentham's own intention that the Panopticon's principles should induce compliant behavior in society. In the midst of Rayner's paean to social media, "Foucault and Social Media: Life in a Virtual Panopticon," he quotes Foucault as saying that "the major effect of the Panopticon [is] to induce in the inmate a state of conscious and permanent visibility that assures the automatic functioning of power." The *automatic functioning of power?* As a starting point, we might ask "whose power, and for what ends?" Raynor also further summarizes what he believes to be the *positive aspects* of life lived intentionally in a social-media-enabled Panopticon:

> There are no guards and no prisoners in Facebook's virtual Panopticon. We are both guards and prisoners, watching and implicitly judging one another as we share content. ... Sharing online is not solely a matter of self-affirmation and self-creation. For many people, the sharing impulse stems from a sincere desire to empower and inform their tribes and communities.[86]

When in all of social or political history has a prison-like environment been considered a proper forum for sharing freely in a communal spirit? Some may protest that Rayner is saying that social media, as we are currently adopting them, are *not* prison-like; a close reading of his essay and a good look at the picture of the prison he placed in the middle of it indicates otherwise. Reading Rayner's essay online, I wondered aloud whether I'd encountered a genuine tech blogger or a shill. He supports inadvertently one of our main theses here, that the way we implement digital technology more often resembles an experiment with live human subjects than it does a reasoned and well-measured attempt to improve our lives.

He opens his essay noting "little is known about how social media is impacting us on a psychological level. A wealth of commentators are exploring how social media is refiguring forms of economic activity, reshaping our institutions, and transforming our social and organizational practices."[87] Yet from such an admittedly shaky starting point—"little is known"—he goes on to cast the entire social media movement in an unbelievably positive light, using words like "empower," "inform," "self-affirmation," and "self-creation" even as he compares current trends to the Panopticon, *to life in a prison.*

If indeed "little is known" about a technology, and your best metaphor to describe it is a participatory Panopticon, then why encourage adoption, even to the point of misrepresenting prison-like conditions as being socially positive? In the end, social media may prove to bring overall positive benefits. Blindly encouraging their proliferation and defending as-yet-unknown impacts without flinching or blanching is the type of behavior that can and does erode genuine digital reliability; powerful but untested technology often harms people—this much nearly all of us can affirm from our own experience (or, for example, by simply watching a documentary on the Manhattan Project). And again, we suggest that one of our criteria for measuring digital reliability is gauging whether we are hurting people in unintended ways with its use.

Returning to Cascio, sousveillance is a critical element of his definition and his discussion about the participatory Panopticon, and it is admittedly an element with apparent potential for doing much good. In fact, Cascio deems sousveillance to have such a bright and powerful future that he again fails to mention that the Panopticon, as Bentham described it, is a prison.[88] (Therefore, Cascio's powerful neologism, i.e., "participatory Panopticon," means literally "participatory *prison.*")

How alarmed then should we be if simply the title of Cascio's essay rings true—i.e., "The Rise of the Participatory Panopticon"? That is, *what if we are* living in an epoch in which the idea of a socially constricting prison is turned on its head, and we all more or less willingly and increasingly use and share digital technology to surveil one another? Fred Reed, in one of his characteristically to-the-point essays, notes, "A problem with strict enforcement of laws by unlimited surveillance is that it will inevitably be misused."[89] This, of course, is an issue of trust.[90]

Certainly when we have the possibility of "unlimited surveillance," we see the issue of trust in a different form and accompanied by profoundly more complex issues and complications than Winner's initial concern that technologies should "work reliably."[91] When the issues of "trust" and unlimited surveillance are involved, our definition of digital reliability actually highlights a clear irony, perhaps even another angle to the Catch-22 described above: the more reliably digital recording devices capture our most intimate and formerly private moments, personal decisions, and actions, the less overall digital reliability we have in the sense that people are being hurt by digital technology—despite the fact that we developed the technology ostensibly in order to help improve the human condition.[92]

An additional problem is that people who are newly reliant upon newly developed technologies are certainly, in many instances, *not* acting in situations in which they should act. For example, the policeman who knows security (i.e., *surveillance*) cameras are sprinkled about a neighborhood likely has far less motivation to walk and work among the people he's entrusted to protect and serve. Again, when we expand properly our understanding of what comprises digital reliability, we find that reliability can, paradoxically, actually be reduced when certain elements of digital technology do the tasks they are built to do; this happens, typically, at the juncture between human volition/action and machine action. We can better comprehend this subtle point when we remind ourselves that every technology, and indeed, every technology-based device, begins with a human-generated plan. Unless we're willing to say that "embracing scientific method" or "employing good industrial processes" ensures perfect plans (i.e., plans that are also executed perfectly), then we are back to our starting point: fallible human will.

In the example above, because surveillance cameras provide a visual window into activities in a formerly human-patrolled neighborhood or community does not mean that all of the benefits of patrolling on foot— e.g., getting to know the citizens, establishing relationships, and (hopefully) building trust—have been replaced adequately by digital technology. We might state more accurately, in fact, that those "side benefits" of patrolling on foot have been *displaced*. On the issue of trust, in particular, we can easily surmise that for many people, seeing cameras placed above their heads makes a distinct statement about a lack of trust, perhaps even engendering a quiet but substantial state-against-the-citizen mentality. This is a way to say, perhaps somewhat simplistically, that we design our digital technology without enough wisdom, without enough forethought, and certainly without clear and adequate comprehension of the point Langdon Winner makes—that today we have "utter dependence upon the large, complex, artificial systems that surround us on every side, giving structure to everything we do."

MIT's Sherry Turkle, exploring in her own way the "intended versus unintended" uses and impacts of robots, generally grants more anthropomorphism to digital technology than we prefer, thereby blurring the lines of human responsibility for how we use our machines. Many of her points are salient to our arguments, nonetheless. While exploring how our nearly unrestrained (and ongoing) acceptance of digitally controlled robots is cheapening our notion of human companionship, she says that "robots are shaping us as well [as we are shaping them]" and even that they are "teaching us how to behave so that they can flourish."

Again, while her points are powerful, we note that since robots did not invent themselves, their "intentions", by definition, come from elsewhere. We do not dispute at all, however, Turkle's notion that we are being collectively conditioned to accept robots (and many other machines) in place of our fellow human beings. In the interest of an effective understanding of digital reliability, however, we emphasize that the intent and fact of societal conditioning comes from the men and women funding, building, and proffering robotics, not the robots themselves. Robots, drones, and various automata are, after all, the soulless objects we ostensibly hope to avoid becoming.

Back to People

Perhaps an even more interesting issue to consider, because it brings us back to what *people* are willingly doing to one another is: What if Cascio (and many other current thinkers and writers) are correct about the *intrinsically good* character of sousveillance; and what if they are also correct in believing that people around the world will rise up, via sousveillance and the technologies upon which it is built, to demand a history-breaking new accountability of both one another and the various levels and forms of government and enforcement that regulate society as a whole?

Morozov explores this question admirably in a chapter named "So Open It Hurts."[93] The participatory Panopticon does nothing if it does not impact society's understanding and—pardon the play on words—exposure to the concept of transparency (or openness). Millions of citizens wielding video-capable cellphones certainly give impetus to the idea that a "Rodney King moment," in which the "keepers" are held accountable via observation (by the *kept*), need not be an isolated incident.[94] Now, much as Cascio predicted, people seem poised to make Google glasses and their tech cousins and derivatives a new ubiquitous tool for watching, recording, and sharing what each of us does, whether acting as citizens, officers, or suspects. As Tom Brignall III notes, "What is unique within the structure of the Internet is that it allows multiple layers of observation to occur such that the 'inmates' can become the observers of other 'inmates'. In such a situation, no one knows who is the observer and who is the observed."[95] We suspect that Google glasses will in no case be the end of personally wielded sousveillance devices; Sony recently announced a patent for wearable "wigs" that their engineers will pack full of sensors and equipment to augment our abilities to digitally sense, control, and surveil within our everyday relationships.[96]

Transparency and openness *are* good, though, in a purported democracy, are they not? A successful democracy or a republic we all (mostly) agree requires informed and knowledgeable citizens no less than it needs wise, aware, and vigilant leaders and government representatives. Before we answer that semi-rhetorical question, we should consider the effect of transparency in everyday life—technology issues aside.

We turn again to Joshua Reeves, who examines transparency, albeit by reviewing life in a different epoch than ours. In his essay, "If You See Something, Say Something," he looks at the history of lateral surveillance in a pre-sovereign historical environment that predates both digital technology and even the concept of the modern police force.[97] Reeves, when he does on occasion touch on modern-day iterations of lateral surveillance in his essay, interestingly eschews analysis of the role of digital technology. He does note, however, that the public (i.e., today's public) has "developed a taste for lateral surveillance" and asserts that while the "new" lateral surveillance we have in the developing participatory Panopticon is "reinforced by a distinct techno-cultural milieu," he indicates there is nothing particularly new about the idea of a government asking for or even demanding that its citizens police themselves via ubiquitous and reciprocal surveillance.[98] For many pundits, Cascio included, perhaps Reeves' account of the Normans' powerful *non-technical* application of lateral surveillance in 11th-century England must be surprising. We don't think of citizens from eras before the industrial revolution as having the free time to watch their neighbors' every move, let alone having the means to inform on them. (Related, at least tangentially, Lawrence Lessig noted that he had more freedom from surveillance, or at least the appearance of such freedom, while touring communist Vietnam than he experiences in the US, in large part because of the difference between the abilities and presence of information systems in the two countries.)[99]

Yet, information systems and technology aside, Reeves chillingly recounts the history of the ancient but very effective *frankpledge* system the Normans employed nearly a thousand years ago, forcing citizens to monitor their own communities and to raise a "hue and cry" the moment they observed criminal activity among their neighbors. This system not only required non-stop vigilance on the part of all citizens, the laws of the Norman leaders also demanded they actually physically chase alleged perpetrators. The penalty for not doing so was a fine for the community and transfer of the criminal's guilt and punishment *to those who observed the crime without dropping their work to pursue the criminal.*[100] In short, we've had forms of lateral surveillance among us long before the advent of any digital technology.

While we are too far removed from the Normans to fully grasp the impact the *frankpledge* system had on the issue of trust among Norman citizens, we can certainly project that a system that forced guilt upon those who did not snitch would not generate a desirable sort of communal trust. A system that makes a virtue out of unfailing blame built upon inescapable lateral surveillance sounds like a community watch designed by demons, like the gates of hell itself. If a goal of the *frankpledge* system was peace and wellbeing, it seems likely the Normans met that goal superficially at best, i.e., via the appearance of public order at the expense of individual virtue, maturity, and the opportunity to exercise wisdom, self-restraint, or mercy in a free environment.

Which brings us to today and the very real rise of the participatory Panopticon in an era bursting with the digital technology to enable it. If others before us have created systems intended to bring order and harmony via lateral surveillance and "responsibilization" of the public, but badly missed the mark in the process, we should consider the reliability of our own attempts to do so, particularly since digital technology so effectively amplifies the ability to surveil.[101]

Transparency, openness, increased accountability—all of these are claimed outcomes of the use of digitally enabled lateral surveillance, sousveillance, and the participatory Panopticon, and many presume each of these outcomes to be intrinsically positive. Returning to our definition of digital reliability, our own starting point in judging these outcomes should be to acknowledge that we should pursue end results that do not hurt others or fail us in unnoticed, unobserved ways, nor should they induce indifference, inactivity, or incapacity where action should be.

Morozov notes the reality that we are prone to confuse the ability to induce an outcome (such as transparency) with the issue of whether we have used technology reliably for good:

> British transparency theorist David Heald draws a useful distinction between transparency as an intrinsic value, as an end in itself, and transparency as an instrumental value, as merely a means to some more important goal, like accountability. Thus, writes Heald, 'the "right" varieties of transparency are valued because they are believed to contribute, for example, to effective, accountable and legitimate government and to promoting fairness in society.' This means, among other things, that there are also 'wrong' varieties of transparency, which might lead to populism, thwart deliberation, and increase discrimination.[102]

Yet, we can rightfully ask how transparency could possibly be negative in a purported democracy or republic, can we not? As a starting point to answer this, Morozov notes that "cognitive science shows that concerns over accountability and transparency greatly affect our decision-making process."[103] In other words, as we mentioned at the outset of this chapter, "the observed" change their behavior, a fact long known in the social sciences and verified by many studies. Scientists can find sub-atomic particles that change their behavior when observed, so certainly it's not incredible to believe sentient citizens would do the same.[104]

Getting more to the core of the issue, we believe an argument can be made that the act of observing, or more specifically, surveilling (i.e., observing with intent to regulate or control), impacts, by default, the issue of trust. Onora O'Neill notes: "increasing transparency can produce a flood of unsorted information and misinformation that provides little but confusion unless it can be sorted and assessed. It may add to uncertainty rather than to trust. ... Transparency can encourage people to be less than honest, so increasing deception and reducing reasons for trust; those that know everything that they say or write is to be made public may massage the truth."[105] We would add, they will almost certainly find their creativity impacted. This point you can prove easily with a thought experiment. Simply note the differences in behavior when the average child plays freely (i.e., one who is unaware of being observed) as compared to when that child is aware of being monitored and held accountable for every behavior and whim.

Mel Gibson explores the issue of trust and transparency beautifully in his 1993 film "The Man Without a Face." Toward the end of that film,

Gibson's character, Justin McLeod, faces a board of inquiry that questions him about his dealings with his young protégé, Chuck Norstadt. The board wants to know if McLeod ensured that the boy's mother was aware and approving of his academic tutoring relationship with Norstadt. McLeod tells the board that the boy assured him he had informed his mother of his meetings with McLeod and about their tutoring relationship, and that the boy confirmed to McLeod that he had gained her permission to continue attending training sessions with him. Given that events subsequently unfolded that make it clear the boy was completely hiding his relationship with McLeod from his mother—leading to (perhaps understandable) uproar and concern in the local community about a potentially inappropriate relationship—the board then asks why McLeod did not contact Chuck's mother to *confirm* that she was aware of and had granted permission for ongoing tutoring sessions.

McLeod's answer to the board of inquiry is telling of our own times; far from cowering at the board's question or apologizing (as most of us would feel compelled to do today), McLeod leans forward and in a semi-confrontational tone "teaches" the board about the nature of trust, making sure they understand that if McLeod (or any instructor or tutor) is to teach effectively, if he is to, in part, ensure the boy learns how to grow to be an honest and virtuous man, then at crucial moments in their relationship he will have to trust the answers that the boy provides about his actions. Why? While unstated in the film, the answer is that we cannot name as "virtue" the actions and activities that flow from awareness of being observed and monitored, nor can we name as "peace" (or even genuine friendship) any apparent harmony that flows from the imposed order of a relationship based upon surveillance.

The movie, based on Isabelle Holland's book, is set in the 1960s. How revealing that Holland was already examining the issue of trust in society and trust in interpersonal relationships, which some would suggest is only now becoming a poignant issue with the advent and implementation of so many technology-enabled surveillance abilities, nearly 50 years ago. Not long after the period for the setting of that movie, of course, the first video cameras appeared in retail stores for the purpose of monitoring our relationships to potentially stolen goods.

Fast forward another 30 years and we have millions of citizens who think it's perfectly natural to drive their cars underneath thousands of expensive taxpayer-funded intersection-monitoring cameras (whose operators and owners remain steadfastly anonymous), citizens who accept anonymous video and digital monitoring of their every communication and action, citizens who accept facial recognition digital technology at Super Bowl games and other public places, and citizens who do not question the intrinsic value and "rightness" of an ever-expanding surveillance- *and* sousveillance-reliant society. This change in such a short period of time reflects remarkably effective acts of social engineering. More important than recognizing that fact is to recognize that we have no idea how such a profound change in our trust relationships (both our trust in relationship to the state and to one another) is affecting our virtue, our maturity, and our innate ability to simply think and act morally and with a modicum of altruism. Actions forced—by physical, psychological, or technical means—rarely engender real moral, civic, or spiritual growth.

Again, acknowledging that people may yet find a way to use sousveillance for overall improvement to our lives, and remembering that forms of state surveillance do have their time and place, the participatory Panopticon itself remains, by definition and without question, a form of prison. This quote from Reeves summarizes our concern:

> [T]he widespread lateral surveillance encouraged by agencies…is alarming because, while the private lateral surveillance…is insidious enough, the organs of a redistributed and increasingly uninhibited policing apparatus are now being plugged into every computer, camera, and other mobile communication device. As policing responsibilities continue to be dispensed to a tech-savvy populace, we should be mindful that these market-driven surveillance technologies will take on an even more pernicious character as the state increasingly relies on technologized citizens to be the eyes and ears of the post-sovereign police....[106]

And Richard Weaver, in *Ideas Have Consequences*, summarizes the problem even more succinctly: "Our planet is falling victim to a rigorism, so that what is done in any remote corner affects - nay, menaces - the whole. Resiliency and tolerance are lost."

We find it difficult to summarize a problem named "control centrism" because of the very real possibility that it is the defining characteristic of our time, meaning that its machinations are all around us and therefore difficult to view or summarize with an objective eye; the political, financial, and military elites who would rule the world seem to embrace control centrism as an inherent virtue in leadership. Popular culture has acknowledged this as far back as the 1960s in the themes of numerous television shows and movies. Buck Henry and Mel Brooks created a popular comedy show named "Get Smart" that featured non-stop interplay between good and bad agencies with the blunt (but nonetheless humorous) names of "CONTROL" and "KAOS."[107] Judging by the show's popularity and various spinoffs it spawned, apparently few American viewers questioned the basic premise of a plot revolving around a bi-polar conflict whose "good pole" was not named "freedom" or "democracy" but— again, ever so bluntly—"CONTROL." This is particularly interesting since the show aired just 20 years after the conclusion of the war that was to restore freedom to the world.

In short, the citizens of our purportedly free republic have long understood, tacitly if not explicitly, that our daily political affairs are dominated not by a drive to extend human freedom and minimize government oversight of citizens—so that they may live, create, and act as a free people—but rather they are driven by men and women who act as if they believe ubiquitous surveillance (with intent to control) is the only framework in which civilization will continue to exist. (Thriving in any aspect of our lives other than the material is also apparently excluded as a goal within this framework.)

The recent refusal from the NSA and the Obama Department of Justice (DOJ) to declassify Foreign Intelligence Surveillance Act (FISA) court rulings underlines the intensity of government belief in control centrism; in the midst of truly global uproar over NSA excesses and baldly revealed NSA and CIA invasions of every form of privacy in every corner of the world, the Obama administration's DOJ gave the FISA court what Mike Masnick calls "the giant middle finger" when the court asked the government to declassify its own ruling on the interpretation of Section 215 of the US Patriot Act.[108] Actions such as these, taken in the current context of global fury against surveillance abuses, reveal the boldness and entrenched power of true believers. Those who stand behind their digital surveillance technology in the face of politically violent criticism from multiple world leaders believe they have an ability *to control,* an ability that the world needs, whether or not citizens *or* leaders understand that need. Whether we nickname this phenomenon a "smothering maternalism," "big brotherism," or any other sobriquet, the fact is that it reflects a control-centric mentality that is powerful enough to resist the pleadings of an entire world full of appalled citizens and leaders. If the original goal of using digital technology to surveil and (ultimately) to control was to keep the world "free," then recent events involving the NSA and CIA reflect, in short, an undeniable form of digital reliability failure.

We cannot with precision describe how leaders of a country that self-describes as "a beacon of freedom for the world" can adopt such belligerent stances in the face of phenomenally broad criticism. Perhaps philosopher Karl Popper, writing just after WWII, summarizes our best guess at the motivations to pan-global surveillance:

> One hears too often the suggestion that some form or other of totalitarianism is inevitable. Many who because of their intelligence and training should be held responsible for what they say, announce that there is no escape from it. They ask us whether we are really naïve enough to believe that democracy can be permanent; whether we do not see that it is just one of the many forms of government that come and go in the course of history. They argue that democracy, in order to fight totalitarianism, is forced to copy its methods and thus to become totalitarian itself. Or they assert that our industrial system cannot continue to function without adopting the methods of collectivist planning, and they infer from the inevitability of a collectivist economic system that the adoption of totalitarian forms of social life is inevitable.[109]

To the extent we embrace "inevitable" digital technology trends that aid the continued growth of the participatory Panopticon, we are perhaps enabling the machinations of those Popper describes as believing "there is no escape" from totalitarianism.

6 HEGELIANISM, DETERMINISM AND THE WHIG THEORY OF HISTORY

The computer's automatic logic, necessary and valuable though it may be, sucks all these flesh-and-blood concerns into a vortex of wonderfully effective calculation—so wonderful and so effective that only what is calculable may survive in our awareness.
The remarkable thing is not this obvious and even hackneyed truth. Rather, it is the fact that, despite our recognition of the truth, we often find it nearly impossible to alter our course in any meaningful way. And so we move in lockstep with an ever more closely woven web of programmed logic. [To get to this place] we had to enter into our own potentials for programmed, automatic thought and action before we could build automatons of silicon, plastic, and metal.

— Stephen L. Talbott, *Remembering Ourselves*

Human memory and recall are famously unreliable, if not downright arbitrary. And the deeper I ventured into aviation, the more critical sharp memory became to success. The array of critical numbers, facts, names, frequencies, and symbols grew with alarming consistency, as did the stress to keep it all straight.

"Checkrides"—this is the innocent-sounding name we gave to the flights pilots undertake with flight examiners to demonstrate their ability to control an aircraft in a wide range of configurations and situations. And checkrides were a too frequent staple of aviation, each flight providing more than ample opportunity for memory failures. These flights often concluded in the debriefing room, with the "oral portion" of the checkride, which always included a colossal test of memory.

Once in the debrief room, the flight examiner pulled out complex aviation charts and pointed to arcane symbols, lines, numbers, and color codes, one after another, asking us to demonstrate that we understood every single detail of our aviation charts and maps. And these details numbered in the hundreds. In particular, we had to know and understand minimum and maximum altitudes on any given portion of a chart. Aviation authorities have dissected and crisscrossed the sky in many places with a mind-numbing array of limitations—including mandatory altitudes for pilots that can make the difference between life and death for any number of reasons.

I remember one checkride in which the check pilot and I sat side by side during the flight portion. My friend Denny followed closely in another plane, with his own checkride pilot seated next to him. We were in the midst of a formation flight checkride, and I knew one particular maneuver we were going to have to demonstrate for the flight examiners was a potential ride-breaker for my friend. In this particular maneuver, I acted as the lead plane, gaining speed and flying up and inverting while my friend Denny tried to stay immediately on my tail. We'd practiced this loop maneuver in formation many times, and during practice, nearly without fail, Denny was unable to keep his plane close enough to mine at the top of the loop, where both aircraft were low on airspeed and therefore less stable than normal. Knowing this particular maneuver was the "long pole in the tent," as they say, Denny and I discussed the maneuver before our formation checkride; we were jointly determined to keep our planes in tight formation throughout.

As our formation check flight unfolded on a steaming-hot July day, we took our planes through their paces, exchanging leads several times to demonstrate to the check pilots that we could act competently as both flight leads and as wingmen. All was going well, but Denny and I both knew the final aerobatic maneuver was the dreaded formation loop, in which we flew in formation tracing an entire loop in the sky, and there was no avoiding it. I was in the lead as we built speed on the "floor" of our maneuver space, then pulled back on the aircraft's stick and pointed my plane up and over, with the sky, the sun, and the light layers of clouds above us flashing in the canopy. I looked over my shoulder throughout most of the loop, keeping sight of Denny. As we neared the top of the loop, inverted, I could see over my shoulder that Denny's plane was nice and close—just where he needed to be. A brief moment of pride flashed through me as I knew I'd kept my lead plane's maneuvers light, moderate, and easy to follow; and, as a result, it looked like my friend and I were on our way to graduation from formation-flight training.

At the very top of the loop, I took my eyes off Denny for the quickest of glances at the altimeter to check my altitude. Any pride I had turned immediately to confusion followed rapidly by a sickening feeling in the pit of my stomach. The loop maneuver we were executing had both a "floor" and a "ceiling" assigned to it—altitudes that we could not go below or above during our aerobatic maneuvers. We started the loop no lower than 10,000 feet—I'd had no problem with the maneuver's "floor"—and at the top of the loop we were to be no higher than, well, *that* was the problem.

For just a few crucial seconds, I couldn't remember our maneuver ceiling altitude. As I pulled my plane down the backside of the loop, with Denny tucked into the formation nice and tight, I remembered the ceiling number and realized I had "busted" my maneuver ceiling at the top of the loop by about 1,000 feet. This was no minor flight violation, as being off a thousand feet of altitude in aviation leads to tragedy—a quick review of aviation history makes that very clear. And worst of all, I had dragged not one but two planes through the busted ceiling. As we returned to the airport with Denny flying close formation on my wing, all I could think of the whole way home was that I had not only "busted" my first checkride due to a momentary lapse of memory, I'd caused my friend Denny to do so as well. That knowledge makes for a long and lousy flight home.

During checkride debrief, my check pilot pulled out his air map. He needn't bother to lecture me about altitude restrictions at that point, though, because my fickle human memory was nice and clear on the ground—sharp as a tack, now that it was too late. We had an assigned "ceiling" of 20,000 feet for the maneuver, and at the apex of the loop I had watched the altimeter tick right up to 21,000 feet, with Denny faithfully in tow. A predictable discussion about the importance of remembering critical altitudes ensued during my checkride debrief. During the entire debriefing, I was as contrite a pilot as the world has known—just waiting for the hammer to fall on both me and Denny.

To my amazement, my flight examiner did not "bust" me that day. He gave me, in fact, a very good grade for the checkride flight, pointing out that I made one very human mistake while I was in the midst of considering—a bit too hard, as it turned out—my wingman and his struggles to stay in tight formation. That flight examiner gave me a lesson in human judgment and human mercy that I won't soon forget. Denny and I met in the hallway after our flight debriefs and high-fived our way down the long and suddenly glowing hallway of our squadron—and I swaggered as only newly graduated pilots can, intermittent memory and all.

Bring in the Digital

A few years later, I flew in a similar high-stress environment, with days full of formation flights, simulated "bad guys," dog fights, and any number of altitude- and airspace-related restrictions. There were planes everywhere in this particular training environment, and among our safety-related training before our first flight, we were shown films of mid-air collisions that had taken the lives of highly skilled and experienced pilots in the same airspace we were using. We were, needless to say, "on our toes," and few of us busted altitude floors or ceilings during that week of training.

If we *had* failed to observe a mandatory altitude—or any other airspace restriction, for that matter—there would be no hiding the mistake in this particular environment. Our aircraft were newly equipped with detail-oriented digital equipment that kept record of our every maneuver and communication throughout the whole flight, for all the world to see. This equipment turned every flight we took that week into a de facto checkride, with all of a checkride's attendant unhelpful stresses, strains, and self-consciousness. I remember thinking back to the formation loop in which I flew a thousand feet too high. With our brand new digital "flight nannies" watching every move, any similar error now would possibly send me packing to the nearest truck driving school (our standard aviator's description for a mistake-induced change of career).

An early reviewer of this chapter of MIAATQ, asked, fairly enough, whether or not we (i.e., the authors) were perhaps confusing ourselves about the value of digital monitoring in aviation. This reviewer's thinking was that maybe a "digital nanny" watching over the shoulders of young pilots was not so bad an idea, particularly in aircraft-intense situations such as military exercises. He said, "I guess the problem I had [when reading this chapter of MIAATQ] was that all of the other personal anecdotes illustrated the potential dangers of digital technology and how our dependence on them can hurt us, but this story seems to show digital technology as being really good. I mean, those altitude warning systems would have prevented Jeff from possibly failing his checkride, and if it were a real flight, could have saved lives." [110]

This reviewer's comment helped me understand that the nuances, depths and complexities of human learning are not always obvious in this day and age, especially when the learning milieu involves a high-end environment such as military or commercial aviation. Checkrides are indeed "real flights," but more to the point, human performance during a checkride may or may not reflect actual aviator performance during emergencies and in high-pressure situations. The type of pressure one experiences from someone "looking over the shoulder" is not the same as the type of pressure one experiences when bullets fly. Viewed from another angle, the type of "situational awareness" one gains from the beeping of digital altitude markers is not the same "SA" one develops from carefully and repeatedly exploring the boundaries of aviation with freedom to make (and self-recover from) human mistakes. In short, the quantified "help" we gain from digitally recording every move and every millisecond of a flight is a devil's bargain: we're so accustomed to quantifying and measuring human behavior in this digital age, that we don't recognize that digital nannies that sound off at every airspace infringement rob pilots of the need to develop their own ongoing caution and self-monitoring. This may appear to some to be a subtle point, but those of us who have been asked to hone the most intricate skillsets involving both intense physical and mental demands understand what we lose when we delegate our carefulness to machines.

While there can be no doubt that digital monitoring of pilot behaviors can help fend off some specific dangerous situations, larger less-visible repercussions ensue when we make permanent certain forms of behavior monitoring. For example, most young drivers experience this phenomenon during the transition from "driving with mom" using a learner's permit to driving on their own with a full driver's license. That first "solo" drive with a new license opens one's eyes to the larger reality of responsible driving. With no one sitting beside you yelping "watch out" every time danger rears its head, the newly free driver learns to internalize life-saving behaviors in a way that cannot be matched in the presence of constant monitoring.

I'd also note that the authors are not trying to paint digital technology as being in and of itself a "bad" thing; we're focused on encouraging people to consider where, when, how and why we embrace all things digital, and we're especially concerned about the countless situations in which we're not even given a chance to weigh for ourselves the merits versus the costs of moving human tasks into the digital realm (Electronic Healthcare Records, anyone?)

How Have We Come to Embrace Flight Nannies (and Other "Thinking Machines")?

We sometimes seem to congratulate ourselves, if not explicitly then at least tacitly, on having created "thinking machines" in our time. Yet an impressive array of writers and world-class experts in the field of digital technology—Weizenbaum, Turkle, Morozov, Talbott, and, more recently, sometime tech-cheerleaders Bill Gates, Elon Musk, and Stephen Hawking—all caution us directly and clearly that machines that replace some of the most human of activities—e.g., *thinking, relating, deciding*—may not, after all, be offering us healthy substitutes for our own efforts in these tasks.[1111] Winner, Postman, and Gatto, albeit not tech writers per se, all present their own implicit concerns on this point as well.

Nicholas Carr, a member of Britannica Encyclopedia's Editorial Board of Advisors, provides a related caution covering much broader turf than just

AI when he notes, "my worry [is] that the Net is sapping us of a form of thinking—concentrated, linear, relaxed, reflective, deep—that I see as central to human identity and, yes, culture."[112] Carr goes even more directly to the point in his seminal article, "Is Google Making Us Stupid?":

> The advantages of having immediate access to such an incredibly rich store of information [as the Internet provides] are many, and they've been widely described and duly applauded. 'The perfect recall of silicon memory,' *Wired*'s Clive Thompson has written, 'can be an enormous boon to thinking.' But that boon comes at a price. As the media theorist Marshall McLuhan pointed out in the 1960s, media are not just passive channels of information. They supply the stuff of thought, but they also shape the process of thought. And what the Net seems to be doing is chipping away my capacity for concentration and contemplation.[113]

In some ways, Carr's concerns are *meta* concerns—not to be dismissed, but concerns that focus on symptoms rather than root causes, and concerns that perhaps fall into the category of "too little, too late." When health experts in multiple first-world nations have identified a growing health concern known as "digital dementia," the cows have likely left the barn, as the saying goes.[114]

So, how *have* we gotten to a point at which digital technology—a technology completely absent from countless successful civilizations in the past—is deemed indispensable for our very existence?

The answer, we suspect, lies not with the power, success, or glamour of digital technologies themselves, as many civilizations before our own have shown an uncanny ability to maneuver and control the material world. People still use roads built by the ancient Romans; even more flock to visit Egyptian and Mayan pyramids (and the building technology they used remains a mystery to us even today). Global navigation, the invention of antibiotics, genetic manipulation (pre-digital, ala Mendel), printing, photography, audio, and even video recording— all predate digital.

Why, then, our current imperative to adopt all things digital, in all places and under all circumstances, seemingly without regard to cost (or the nature and side-effects of the cost)? People in our time seem to embrace and even immerse themselves in the very concept of "technological determinism," and to do so with a particularly religious fervor. For an example of a high priest-like rebuttal to dissension among the digital elite ranks, read Clay Shirkey's well-written (but almost obscenely worded) objection to Carr's "Is Google Making Us

Stupid?"[115]

We may not, in our time, ever truly understand the motivations toward all things digital. That does not mean, however, that we are blind to the concepts, maneuvers, and tactics some use to deepen and further the acceptance of all things digital.

One unmistakable tool (or tactic?) in use by governments, corporations, and just about any organization large enough to influence public opinion and policy is the use of Hegelianism as a means to address technology-related problems. Hegel and his works (and thousands of derivatives from his works) cannot be covered in depth here. We do note, however, that many of those with a deep understanding of Hegelianism and its roots in philosophy use one of Hegel's basic concepts, triadic development, for the further adoption of digital technology. Sadly, perhaps many more folks employ aspects of Hegelianism without possessing the slightest understanding of its philosophical roots and implications, and this unfortunate fact is fairly plain to see, simply by following how policy moves and operates in our nation.

Again, in hopes of keeping our discussion relatively simple, we begin with a description of Hegel's thought from New World Encyclopedia:

> [Hegel synthesized conflicting religious and cultural elements] with the radical concept that, contrary to Aristotle's portrayal of the nature of being as static and constant, all being is constantly in motion and constantly developing through a three-stage process of **thesis, antithesis**, and **synthesis**.

This theory of **triadic development** *(Entwicklung)* was applied to every aspect of existence, with the hope that philosophy would not contradict experience, but provide an ultimately true explanation for all the data collected through experience. For example, in order to know what liberty is, we take that concept where we first find it, in the unrestrained action of the savage, who does not feel the need to repress any thought, feeling, or tendency to act. Next, we find that, in order to co-exist with other people, the savage has given up this freedom in exchange for its opposite, the restraint of civilization and law, which he now regards as tyranny. Finally, in the citizen under the rule of law, we find the third stage of development, liberty in a higher and a fuller sense than that in which the savage possessed it, the liberty to do and to say and to think many things which were beyond the power of the savage. In this triadic process, the second stage is the direct opposite, the annihilation, or at least the sublimation, of the first; and the third stage is the first returned to itself in a higher, truer, richer, and fuller form.[116]

The pattern described in this simple example—triadic process in action in a story about a hypothetical savage—is, strangely enough, evident with even a casual look at how our government and corporations repeatedly promote ever-increasing use of digital technology in our lives. The following excerpt from a *New York Times* article about online security describes triadic development in action in the clearest of terms, and the parallels to the savage/slave/citizen progression described above are plain to see:

> Although technology companies say they take security seriously, protecting their customers seems to come second to announcing new products. Take Twitter, where people's accounts are frequently hacked. In the last few months alone, this has happened to Burger King, BBC, NPR, The Associated Press and a slew of celebrities and users. In that time, Twitter has proudly announced updates to features on its mobile and desktop apps, introduced a music Web site and redesigned its company blog. But it still hasn't released two-factor authentication, a security tool used by the rest of the industry to deter hackers.
>
> One solution is a bill crawling through Congress over the last two years, the Cyber Intelligence Sharing and Protection Act, known as Cispa. The bill would make it easy for tech companies to share information about computer security threats with government agencies, helping fortify against cyberattacks.
>
> But privacy groups say that *Cispa is not a solution to the problem, and that instead it hands the highly sensitive personal data we want protected to the government.* 'It has to be the obligation of these tech companies to build in security from the very beginning before we start moving into solutions about bringing the government into the private sector,' said Leslie Harris, president and chief executive of the Center for Democracy and Technology, a Washington-based advocacy group financed by a broad coalition of technology and telecommunication companies.[117] [Emphasis mine.]

This excerpt describes clearly a triadic progression: the government wants (and in some fields, demands) more use of digital technology in our lives;[118] hackers are upsetting citizens by accessing and exposing formerly private personal data now held in (newly vulnerable) digital forms; And then the government—working hand in glove with the digital technology industry—introduces legislation to enable more sharing of personal data with the government in order to stop the hackers.

And this process is nearly always the same: digital technology solves some existing problem (sometimes in spectacular fashion), but in the process it introduces one or more new problems. Some sector of society then notes the new problems and asks for a solution. Government and/or industry suggests a solution involving more digital technology, a solution coming, all too often, from the same product, company, or sub-industry that generated the new problem in the first place.[119]

This may sound like a simplistic example—and in fact it is a technique that governments and corporations find very simple to execute—yet it represents a technique that governments, corporations, non-governmental organizations (NGOs), and well-funded foundations use, without exaggeration, on a daily basis with the deepest of impacts on our lives. Does the sequence of actions described above happen "by accident" at times? Surely. Are there extremely educated individuals, schooled in human psychology and sociology purposefully employing those techniques in attempts to manipulate and control society? Just as sure, the answer is "yes," unless we are willing to dismiss many of our national actions and interests as they played out throughout the 20th century and into the present.[120]

One telltale mark of Hegelianism at work in our society—in fact, we can think of this as "the smoking gun" or the dead canary in the mine shaft—is that alternatives to the third part of the triadic process, the "synthesis," are rarely acknowledged, nor discussed with any serious intent to implement.

In the example above, for instance, a number of viable alternatives to the Internet security issue come to mind. Some alternatives to Cispa involve other types of government actions (of course). Congress could hold hearings to consider if commercial companies are giving proper attention to the security of personal data. Congress could, perhaps, even consider adopting the European model of protecting people's personal data—a policy widely disparaged in this country but one that grants far more dignity and privacy to European citizens than our laws grant us. (Yes, privacy and dignity are eternally and irrevocably linked.) Congress could also enact, or at least threaten to enact, laws providing strict penalties for breaches of security involving private data online. Or they could even begin a serious review of just what we are losing as a society in our head-long rush to digitize every aspect of our lives. The list of alternatives to Cispa's legalization of additional personal data transfers to the government is as long as it is (apparently) invisible to the majority of our media.

Even more important, perhaps, are the alternatives we have as "victims" of this latest round of Hegelianism. We can, for now at least, opt to *not use* the sites and companies that fail to protect our personal data. What might happen, for instance, if a significant percentage of Twitter users contacted the company to let them know that without two-factor authentication in place, Twitter's services would no longer be required?[121] And we always have available the ages-old (if lumbering) process of contacting congressmen to let them know how important we consider our online privacy. (Even a blind squirrel finds the occasional acorn.)

Note also the "obvious" statement by Leslie Harris in the quotation above: "It has to be the obligation of these tech companies to build in security from the very beginning before we start moving into solutions about bringing the government into the private sector." As to why technology companies do not always build the obvious into their products from the outset, the very success of Hegelianism perhaps provides the answer: why build perfect products and sell them once or twice to the public when you can build interesting but fundamentally flawed products and sell the maintenance, support, and upgrades for them continuously? Before screaming "conspiracy monger," the reader should recognize we just described a variation of planned obsolescence, a tactic used by the automobile industry that has been well-known and well-documented since the 1950s. If you think that method for generating constant cash flow remained the private business tactic of builders of cars, read the article "Planned Obsolescence" in the March 23, 2009 edition of the Economist, which bluntly states, "The strategy of planned obsolescence is common in the computer industry too."[122]

We too often seem to allow Hegelianism to negate our ability to see and consider alternatives, such as walking away from vendors or industries that scorn reliability for the sake of obtaining ongoing business. In fact, for those intent upon manipulation, our pliant reactions to Hegelian-based approaches may be the raison d'etre for Hegelianism itself. Why we seem so powerless against the manipulations of this well-worn and nearly ubiquitous variation of social engineering seems at times a mystery. Perhaps its pairing with the Whig theory of history, a marriage at times that seems to provide the default set of historical lenses for the entire English-speaking world, renders us incapable of recognizing and understanding problems that *should not* be solved by "technological process." In other words, we perhaps feel that any action rendered in remediation of a perceived technical problem must come from technology itself.

And when anything resembling a Gordian knot solution does appear (e.g., setting down our texting devices and walking away), our incredulous reactions quickly reveal such solutions as default non-starters. We are not prepared, it seems, to discard or abandon any technology once it is in place, no matter how destructive (Hiroshima), antithetical to good (digital dementia), or even deadly (Chernobyl and Fukushima) our experience with it proves to be.

Of course, sometimes the best way to avoid the problem of a Gordian knot is through exercising caution *before* someone creates the knot. Jeffrey Brady, of National Public Radio, recently completed a journalistic piece about the use of technology among the Amish. As he interviewed various Amish folks, he made clear the fact that the Amish *are not unremittingly against technology.* Their approach to technology, however, covers one of our own core considerations when judging the reliability of technology: The Amish proceed with caution before adopting a technology so that *as a community, they can judge in advance* the likelihood of [technology] "failing us in unnoticed, unobserved ways." Brady notes that while most of us automatically consider the newest technology to be a good thing, "The Amish don't automatically embrace what's new. They evaluate it and decide if it's a good fit for the lives they want to lead ... [T]hat is where the Amish may have a little something to teach the rest of us."[123] There is of course a simple formula here:

> A equals any given new technology.
> B equals Amish belief they have a choice about adopting and accepting any given new technology.
> C equals either subtracting that new technology from their lives or adding it.

This simple approach contrasts starkly against the "new math" of Hegelianism that goes invariably from A to C without the pesky admission that we *always* have the choice of rejecting a given technology.

Richard M. Weaver, writing more generally about materialism and positivism—which make up the soil in which our tech-centrism flourishes—may have best captured the feeling of our dilemma when he said that modern man "has found less and less ground for authority at the same time he thought he was setting himself up as the center of authority in the universe [via his reliance on tools and techniques that manipulate the material world]; indeed, there seems to exist here *a dialectic process* which takes away his power in proportion as he demonstrates that his independence entitles him to power."[124] [Emphasis mine.]

Let's reconsider, hypothetically, of course, the idea of simply walking away from our technology-based devices. Many professionals, even those within the computer industry itself, have found solace in the occasional hike, mountain climb, or camping adventure in which they intentionally abandon all devices and networks that could tie them back to their tech-immersed lives. We often hear their sighs of satisfaction when they recount the liberation they experienced during these self-designed reprieves. Most often, though, they come back to society and resume their tech-laden lives as if never missing a beat.[125]

On the surface, this is seemingly contradictory and nearly arbitrary behavior. Rational behavior, it seems, would suggest that upon finding a way of life that feels deeply liberated compared to our current experience, we would make certain to begin the process of abandoning the chains that bind our experience of "liberation." Yet, people rarely seem willing to do this when technology is involved, even after having experienced self-described exhilaration upon walking away from tech-enabled chains.

The beauty of simply walking away from a given trouble-making technology is that in taking this action, we break—irreparably—the Hegelian dialectic; Hegel's triadic development depends upon an initial "problem"—the thesis. This initial problem (which, somewhat ironically, must remain in place for Hegel's approach to work) leads to the antithesis, the pain of which is of course necessary in Hegelian thinking for motivating people to reach (or accept) the synthesis. For those who would manipulate people by using this process in support of hidden agendas—and these persons seemingly are legion—part of the strength of this process is that it "retains" aspects of the thesis, the original problem. So if leaders with political agendas see any given technology as critical to their future plans for society, the Hegelian approach provides the perfect means to "problem solve" while keeping in place (or at an absolute minimum, "in view") pet technologies. Our walking away from any given problem-technology *breaks permanently* any opportunity to preserve that technology within a Hegelian synthesis. Walking away, stated simply, is kryptonite to Hegel.

The Great Stereopticon Meets the Participatory Panopticon

As for answering why we don't break the Hegelian chains following a refreshing hiatus from techno-centrism, we turn again to Richard Weaver for an explanation of the typical self-directed "return to chains." Weaver, writing during the heyday of press, radio, and pre-Internet television, and just before we collectively embraced digital technology, argued in his classic book, *Ideas Have Consequences*, that the pervasive effects of mass media were being used to contain and condition the masses. This inherently manipulative use of mass media, which he labeled the "Great Stereopticon" was meant to keep everyone as firmly and permanently as possible in their bourgeois state:

> [It] might appear that this machine [i.e., the Stereopticon], with its power to make the entire environment rhetorical, is a heaven-sent answer to our needs. We do not in the final reckoning desire un-interpreted data; it is precisely the interpretation which holds our interest. But the great fault is that data, as it passes through the machine, takes its significance from a sickly metaphysical dream. The ultimate source of evaluation *ceases to be the dream of beauty and truth and becomes that of psychopathic, of fragmentation, of disharmony and nonbeing.* The operators of the Stereopticon by their very selection of matter make horrifying assumptions about reality. For its audience that overarching dome [of the Stereopticon] becomes a sort of miasmic cloud, a breeder of strife and degradation and of the subhuman. ... *It is the work, too, of many who profess higher ideas but who cannot see where their assumptions lead.* Fundamental to the dream, of course, is the idea of progress, with its postulate of the endlessness of becoming. The habit of judging all things by their departure from the things of yesterday is reflected in most journalistic interpretation. Hence the restlessness and the criteria of magnitude and popularity.[126] [Emphasis mine.]

"Fundamental to the dream, of course, is the idea of progress"; *in that phrase Weaver identifies the Whig theory of history clearly* as one of the sources of much of what he calls the "sickly metaphysical dream." He also reiterates a foundational belief of Hegelianism—that of the "endlessness of becoming." This concept, which is directly contradictory to the traditional Christian belief that the source of being has a never-changing essence, interestingly, is key to both Hegelianism and the Whig theory of history. Thus, our erstwhile understanding of the permanent essence of the Creator, which is *prevalent in much of Western thought for the last 2000 years*, is, according to Weaver, flaunted and ignored by our current intellectual movers and shakers.

How interesting that Weaver wrote these thoughts, describing a *Nachtmahr* of a technology-enabled "Stereopticon," just before we enabled a fully operational participatory Panopticon via ubiquitous digital technology. And how tempting for true believers in digital technology—Morozov's "solutionists' and "Internet-centrists"—to believe the increased "interactivity" and connectivity the Internet provides (as compared to its closest cousin, television) will somehow break the dome of Weaver's Stereopticon.

The fact that reveals the Emperor's (i.e., digital technology's) true state of undress, however, comes through in the way Weaver built his analysis and condemnation of the world of controlled mass media and mass communication. He searched, of course, as the name of his book implies, for *ideas* that were to blame, not focusing solely on specific communication technologies or particular subsets of media. He, in fact, quite self-consciously (and conspicuously) began his *written* arguments with Socrates's ancient critique of the very medium Weaver uses to make his own arguments—the written word itself. He both acknowledges and forestalls the potential criticism of using a medium hypocritically, i.e., by simultaneously using it and recognizing ancient criticisms of it, by noting that we can use technologies wisely (e.g., such as the printed word) when we make due reference and consideration to a specific technology's purpose and circumstances.[127]

Weaver's concern about the Great Stereopticon, which he voices clearly in the chapter by the same name in *Ideas Have Consequences*, can be found in some of the following excerpts:

> The problem...in the lap of practical men, those in charge of states, of institutions, of businesses, [i.e., in our terminology, today's *movers and shakers*] is how to *persuade to communal activity* people who no longer have the same ideas about the most fundamental things. In an age of shared belief, this problem does not exist.... [But the] politicians and businessmen [of today] are not interested in saving souls...they are interested in preserving a minimum of organization, for upon that depend their posts and their incomes. [Emphasis mine.]

How subtly Weaver names our apparent sin du jour, love of control, in his restrained mention of the drive for a modicum of "organization." Is he, taken in context, perhaps suggesting that the pre-digital electronic age served as a precursor, or even provided a substructure, for the completely monitored and controlled interpersonal communication we have today? Yet free *communal activity* is what the Internet and social media are all about, right? Or maybe it's about education, or access to information, data availability, transparency, and transformation? Or—what if the Stereopticon Richard M. Weaver saw and described in the 1940s has neither gone away nor become irrelevant to our secular state, but instead is coming to fulfillment in Internet-based social media?

What, in fact, if Hegelian triadic development brought us incrementally from the crystal-based radio sets and family Saturday evenings around the RCA of the 1920's to the NSA-monitored telephone calls, email exchanges, and recorded social media sessions of our current participatory Panopticon? Today we have the combination of a populace quietly but completely embracing the Whig theory of history, i.e., a whole nation full of people believing that in fact, and without fail, everything, technology especially, improves day by day (and indeed cannot help but continue to improve), and that same population routinely subjected to Hegelian-based social engineering.

This provides a powerful one-two punch that keeps the heads of even the most inquisitive would-be philosophers and pundits low and quiet—the kind of dangerous folk Weaver warns about who can observe manipulation, political or otherwise, as it unfolds and who can perhaps share (dangerously) some society-challenging wisdom with their peers.

7 SOCIAL ENGINEERING

Modern technology has become a total phenomenon for civilization, the defining force of a new social order in which efficiency is no longer an option but a necessity imposed on all human activity.

— Jacques Ellul, (n.d.). BrainyQuote.com. Retrieved June 14, 2014[128]

Too often we hear it suggested that some form or other of totalitarianism is inevitable. Many who because of their intelligence and training should be held responsible for what they say, announce that there is no escape from it … In fact, one should not enter into a discussion of these specious arguments before having considered the following question of method: Is it within the power of any social science to make such sweeping historical prophecies? Can we expect to get more than the irresponsible reply of the soothsayer if we ask a man what the future has in store for mankind?

— Karl Popper, *The Open Society and Its Enemies*[129]

For hundreds of years the picture that has emerged from science is that we're basically made from material stuff and we work in mechanistic ways. … Here's a metaphor that explains the whole thing. Assuming that I'm a decent mechanic and I know how cars work, I can take my car which is running, turn it off, spread the pieces all over the driveway, put it back together, turn the switch, and it'll turn back on. If I take my dog and cut him up into little pieces and spread him all over the driveway, and I'm a really competent surgeon and I know how to reattach all those pieces, and I put 'em all back together, he's not gonna' bark at me. There's something fundamentally different between machines and life. And we are running our society as if we are pieces of a machine, and as if the world is a machine.

— Dean Radin, Senior Scientist, Institute of Noetic Sciences and Thom Harmann, author of *Last Hours of Ancient Sunlight*, quoted in *I Am*, a reflective documentary film by director Tom Shadyac

Look at Me, Please!

Millions right now are sharing what they're doing, how they're feeling, and what movie they're at, all with a few clicks or taps on their mobile devices. Most of the sharing is voluntary, and the users are (mostly) in control of what is being shared. However, the companies that provide the places to store personal photos and private thoughts are becoming more willing to decide for us what should be shared. With smart homes, smart cars, and smart appliances—enabled by what's currently being called the Internet of Things (IOT)—we are on the verge of creating enormous amounts of data that can be "mined." Our massive migration to social media as our default means of communicating with one another is just the warm up. When social media started, it was exciting to broadcast quips, jokes, thoughts, or troubles to a group of friends. We no longer had to know how to build a website to let our voices be heard or to express ourselves with artsy photos and writings. Yet, as many folks using these services can attest, with millions and eventually billions of users enrolled, we find continually expanding user agreements and settings changes that impact what information any given application shares automatically. There may be a few weeks of backlash when companies introduce these impactful changes, and back-and-forth exchanges revolving around that tired bromide, "If you don't want to use the service, then just cancel." But, many (especially younger people) have grown accustomed to relying on these services to keep in touch or to be part of "the conversation."

One example of automatic sharing can be seen in Spotify, a music streaming service, that integrates itself with other social media networks and instantly displays what a user is listening to on a shared "wall." I, for one, used to hide my iPod screen in embarrassment, if there was a chance someone could see what I was listening to while on a plane or as I walked through a store. But, in a few short years, we have gone from worrying about someone catching a glimpse of our play screen, to services tracking every song we listen to, when we listen, the frequency, and where we are when listening to them. These services also know which of your *friends* like the same song. Then there are the GPS-running apps that can show you where you ran, provide your pace, and can compare all the data from previous runs. You can even share (i.e. post to your social sites) your running route and time for all to ogle. After a recent jog, I received an email that showed I ran my 3.6 miles in the top 70th percentile for speed-to-finish, with a custom graphic of a police man giving me a "speeding ticket." I have to admit, I momentarily swelled with pride upon receiving this glib email. Having given it a bit more thought, I now recognize the inherent creepiness of this new world. Why is my route, my jogging speed, and the "percentile" in which my performance fits a fair topic for permanent storage on some unknown server, let alone a candidate for public broadcast? The manner and ease with which corporations and social media services (Instagram, Facebook, Twitter, etc.) continually expand what it is we share, with little backlash, indicates highly effective social conditioning.

With now-ubiquitous social media, we're conditioned to accept an inherently non-private means of communicating, and many of us are even willing to support directly the infrastructure that enables all of this. Apple finally got right to the point starting in 2015: you can now pay 32 dollars per month for a yearly iPhone upgrade program. The unstated rationale behind this is "Why would anyone nowadays expect a phone to be useful for more than a year?" Companies have induced massive changes in our expectations for the durability and reliability of their products and services. This is also a form of social engineering; encouraging us to accept constant fluid change to the most basic elements of our day-to-day lives makes it far easier for companies (and, if we're being honest, *the governments in bed with those companies*) to use incrementalism for achieving goals that are based in ulterior motives. The same phone that needs annual replacement, keep in mind, will soon be connected to your home's thermostat, front-door lock, lights, and your fitness routine.[130]

Many aspects of life have become more convenient, and we certainly know a lot more *trivia* about one another than any other generation. But, at what cost? Where is this heading? What are the risks, and who, if anyone, is calculating them?

Could it be that things are simply moving too fast? Simply trying to research and collate the huge number of stories related to privacy, drones, social media, and robotics could be a full-time job. But we also sense a general and understandable apathy out there—like the feeling you get when clicking "I agree" on a 400-page iApp agreement. Whether or not this feeling is intentionally induced by Madison Avenue, there will be those who take advantage of it now that it's in place. Let's take a look at where we're headed through the lens of some of the great thinkers of the past.

Relevance for Digital Reliability

Karl Popper, in his masterpiece *The Open Society and Its Enemies*,[131] discusses a wide range of authors and ideas related to the subject of societal "openness," covering literally thousands of years of thinking and writing on the subject. Interestingly, the subtitle of the original release of Volume I of Popper's book is "The Spell of Plato." Few of us in the West think of Plato in negative terms, yet Popper shows masterfully, through many direct quotes, that Plato's teachings about the role of the state in relation to the "free" individual seems to have influenced deeply the thinking and actions of many of the most repressive of 20th century's autocrats and tyrants.

Popper's massive work extends beyond the scope of this book, but he focuses very early in his work on the simple (but potentially dangerous) concept of *collectivism*. He wrote the book during the late stages of World War II and freely admits he allowed the trauma and terror of that conflict to impact his writing. This may be why the very first endnote of his book discusses what Popper deems to be a pillar of the modern repressive state—a type of collectivism that focuses not on a proper and healthy "rational institutional planning for freedom" but one that instead involves what Popper calls "group egoism." Popper illustrates the critical difference between these two flavors of collectivism. He distinguishes between using God-given rationality in planning for and maintaining commonly held freedom and prosperity, as opposed to the marshalling of masses for the purposes of suppression and control. He quotes Charles Sherrington to illustrate the vast difference between those two types of collectivism: "Are the shoal and herd altruism?"[132]

In quoting Sherrington, Popper notes succinctly that simply "doing things together" does not magically invoke "the wisdom of the crowd" (if such a thing exists). Sherry Turkle, in *Alone Together*, goes a step further and points out that many of today's digitally enabled "group activities" don't even enhance human communication, let alone induce any sort of collective wisdom. Even the title of Turkle's book speaks to the pseudo-communities we willingly adopt today, as we gladly sacrifice yesterday's social units, face-to-face gatherings, and even age-old family activities in the name of a digitally enabled form of communication—the brevity and warmth of which resembles machine language in too many ways to count.

False promises of today's legion of techno-cheerleaders aside, Popper notes that of course we frequently have a need to act collectively, and in his writings, he does not even hint at deprecating social institutions, viewed as a whole.[133] His real concern is that he, and indeed the entire world, watched tyrants of the 20th century—armed with new technologies in communications, military arts, and social control—march millions of human beings into political slavery and submission to the state. If today we fail to understand our own mentality (and, of course, we include the mentality of our leaders here), then the possibility of new forms of tyranny waits in the wings, a tyranny newly empowered by the latest digital and nano technologies.

Were Popper alive today, it's fair to ask whether his *major* concern would be the danger of individual leaders abusing political power. Given that Popper started his arguments by discussing collectivism, not individual tyrants, and given that he made a very clear distinction between the type of democratically led collectivism that lends itself to the maintenance of human freedom, as opposed to those types of collectivism that dull and enslave, we can safely say that the focus of Popper's concerns aligns with those of Alexis de Tocqueville.

De Tocqueville wrote in the 19th century, and he made clear that his greatest fears about America's future (i.e., *today*) were in reference to the American masses. De Tocqueville feared the mistakes *we*, the everyday people, are prone to make more than he feared any other future political eventualities—even the actions of individual tyrants. His accuracy of prediction about future political developments in America alone are noteworthy, and de Tocqueville's work in general is well known and widely quoted. That's why it's all the more surprising and painful that, Oedipus-like, so many of us continue to walk the dangerous political paths he warned about so clearly, so many decades ago.

To illustrate, we'll begin with select quotations from de Tocqueville's *Democracy in America* and compare and contrast his projected concerns—points of caution he uttered in the early 1800s, no less—with some political/social commentary published recently by our mainstream media:

> I think, then, that the species of oppression by which democratic nations are menaced is unlike anything which ever before existed in the world: our contemporaries will find no prototype of it in their memories. I seek in vain for an expression which will accurately convey the whole of the idea I have formed of it, the old words despotism and tyranny are inappropriate: the thing itself is new, and since I cannot name, I must attempt to define it. ... The first thing that strikes the observation is an innumerable multitude of men, all equal and alike, incessantly endeavoring to procure the petty and paltry pleasures with which they glut their lives. Each of them, living apart, is as a stranger to the fate of all the rest–his children and his private friends constitute to him the whole of mankind; as for the rest of his fellow-citizens, he is close to them, but he sees them not; he touches them, but he feels them not; he exists but in himself and for himself alone; ... Above this race of men stands an immense and tutelary power, which takes upon itself alone to secure their gratifications, and to watch over their gate. That power is absolute, minute, regular, provident, and mild. It would be like the authority of a parent, if, like that authority, its object was to prepare men for manhood; but it seeks, on the contrary, to keep them in perpetual childhood ... For their happiness such a government willingly labors, but it chooses to be the sole agent and the only arbiter of that happiness; it provides for their security, foresees and supplies their necessities, facilitates their

pleasures, manages their principal concerns, directs their industry, regulates the descent of property, and subdivides their inheritances: what remains, but to spare them all the care of thinking and all the trouble of living?[134]

Keep in mind that de Tocqueville penned these words before the American Civil War. Remember also that his works are widely published, famous, and (currently) freely available. And with those thoughts in mind, read the following select quotations from an opinion piece, penned by Mike Weston and entitled "'Smart Cities' Will Know Everything About You," published in 2015 by the *Wall Street Journal*:

> From Boston to Beijing, municipalities and governments across the world are pledging billions to create 'smart cities'—urban areas covered with Internet-connected devices that control citywide systems, such as transit, and collect data. Although the details can vary, the basic goal is to create super-efficient infrastructure, aid urban planning and *improve the well-being of the populace.*

> … In a fully 'smart' city, every movement an individual makes can be tracked. The data will reveal where she works, how she commutes, her shopping habits, places she visits and her proximity to other people. … this data will be centralized and easy to access.

> Imagine the scenario: A beverage company knows a particular individual's Friday or Saturday night routine. The company knows what he drinks, when he drinks, who he drinks with and where he goes. It also knows how the weather affects what beverage the individual chooses and how changes in work patterns influence how much alcohol he consumes. By combining this information with the individual's social-media profile, the company could send marketing messages to the person when he is most susceptible to the suggestion to buy a drink.

> A smart city doesn't have to be as Orwellian as it sounds. If businesses act responsibly, *there is no reason why what sounds intrusive in the abstract can't revolutionize the way people live for the better by offering services that anticipates their needs* [sic].[135] [Emphasis mine.]

Reading Mr. Weston's outline of a "tech utopia," we're struck by how unaware he appears to be of the social and political thought history that lay behind him. If he's less ignorant of that history than we fear, then we're presented with a far more worrisome possibility: It's as if he took de

Tocqueville's words of warning and used them as an outline for the "tech utopia" he describes.

Weston, as he indicates, is by no means alone in his belief that ubiquitous monitoring of the non-criminal populace is somehow a good thing. The men and women acting "in municipalities and governments across the world" to fund these all-knowing "smart" cities could be acting, as Weston says, to "improve the well-being of the populace," or there could be other motivations at work. In either case, multiple government leaders in disparate locations around the world don't typically spend "billions" to radically modify infrastructures without sharing some unifying or common motivation. Can it be that leaders believe the dystopia of de Tocqueville's brilliantly accurate prediction is really a misunderstood form of utopia? Or, far more disconcerting, could it be that some portion of our leaders embracing "smart cities" deems the modern masses *worthy* of a Panopticon living environment? Which is to say they perhaps deem us *un*worthy and/or unprepared to rise above, as free men and women, the "immense and tutelary power" of de Tocquesville's worst fears.

Motivations aside, one of the most fascinating aspects of this whole affair is the audacity (or elite insight?) of those writing and publishing such opinion pieces. As we noted above, both Orwell and Huxley are known throughout the world for having written quite plausible dystopic novels, both centered around social engineering of the masses. And in both novels, that social engineering relied, first and foremost, on a massively intrusive state. People took Orwell's work so seriously that they formed at least one organization dedicated to ensuring his dystopia remained *only* a literary fantasy. In other words, in regard to the risk of America converting itself into a massive "nanny state," de Tocqueville was by no means a lone voice crying in the wilderness.

This brings us full circle to considering, again, the role of digital technology in our society. The early 20th-century writings of people like Bernays and Spengler and the actions of states such as East Germany in the mid-to-late 20th century provide clear evidence, long before the advent of the Internet and cellphone cameras, of the will to both surveil and engineer (i.e., manipulate) entire populations. Against both of these groups (i.e., both the writers and the implementers), we watched many people, and indeed, entire blocs of countries, rise up in alarm and even armed conflict. The Stasi and their repressive and intrusive antics against their own people were nearly as familiar to Americans in the 60s and 70s as the casts of the police-centric television shows *Adam-12* or *Dragnet*.

One wonders what a Bernays, whose concepts about moving the masses via propaganda were not lost on Nazi Germany, would have done with the type of broad-spectrum databases of the "smart cities" Weston describes? Far more salient, though, is the question of how we've come to the point that national media can carry articles about "all-knowing smart cities" without a corresponding public outrage? We are post-Nazi, post-Stasi, post-*1984*, and even post-Snowden. We have to wonder, then, what happened to the outrage, shock, and seeming abundance of caution people carried in hopes of disabling or mitigating the mentalities and structures of the silently engineered state? In 1989, as the Berlin Wall came down, along with it came unfathomable stories of Stasi tactics that turned family after family against one another in the name of social engineering, *in the name of maintaining social stability and control.* Imagine the world's reaction if at the same time the Berlin Wall fell in 1989, the *Wall Street Journal* published an article entitled "'Smart Cities' Will Know Everything About You." That would have been a "from the frying pan into the fire" moment if ever there was one.

So how is it that so many folks appear to welcome the fire now? What factors, processes, bad habits, or, frankly, *technologies* brought us on the long, long social journey from shock and disgust to willing acceptance, longanimity, and even complicity? In less than a century, we went from hoping for a "chicken in every pot" to a "camera in every corner." And again, we made that journey with a brilliantly visible backdrop of horrific examples of life (and death) within real-life nanny states. Those of us who visited the escape museum at Checkpoint Charlie at the Berlin Wall in its heyday know the outrage and active resistance against all-knowing states was—not so long ago—very real, extremely intense, and well documented. Part of the answer to "How did we get here so fast?" might be that we've been on the journey longer than we suppose.

Wherever we are on that journey, our current stop should be labeled "extremism." To wit, current proponents of social engineering even embrace the literal genetic engineering of individual people within society, sometimes seemingly for the sake of the concept of society alone, soteriology be damned; at least one well-informed proponent of the "metaman" or "post-human" concept openly talks about the various strata of future societies he believes will ensue, full of genetically engineered people. And he apparently accepts, without blanching, the prediction that we will one day soon engineer human "workers" who will be so far beneath the engineered elite humans that cross-procreation between the classes will cease.[136] This sort of future, which many geneticists and artificial intelligence researchers *are working right now to implement,* certainly appears to go far beyond the primitive forms of eugenics promoted in the early 20th century by American thinkers—eugenics which, sadly, were adopted and taken to even more horrible extremes in Nazi Germany.

How can we have forgotten already the recent tragic lessons of real-life eugenics? The answer to that may lie in the fact that while we still cringe at (true) stories of forced sterilization in American public schools, and we continue to build monuments and museums in remembrance of the Jews, priests, homosexuals, and other unwanted masses the Nazis and Marxists attempted to "engineer out of existence" during the WW II era, *the underlying philosophy that forms the absolutely essential foundation of social engineering has never been repudiated*. Its roots are in the still-revered philosophy of a madman (Nietzsche), and according to early 20th century writers such as Spengler, those philosophical roots extend even further back to the much-celebrated "Enlightenment" period.

Spengler, Nietzsche, and an Unrecognized Form of Engineering

Setting aside, for the moment, the extremes of some of the proponents of "metaman" and "post-humanism," let's consider some of Spengler's thoughts, if for no other reason than the interesting timing of when he wrote his *Man and Technics (Der Mensch und die Technik)*, which was 1931. At that time, we in America were already well into our own eugenics movement, as both academia and the public had had time to digest Darwin's ideas and Mendel's work; during this period of the last century, newly aware of the secrets of genetics, we Americans collectively had more to say (and do) about keeping "good genes" among us than most of us would like to remember. That is to say, the soil for growing social engineering in America was rich.

Spengler's *Man and Technics* was a smaller and less well known than his magnum opus, *The Decline of the West*. Nonetheless, his topic provided a much-neglected look at technology and its impact on society, and within the confines of that small book he struck a sort of balance in that he likely insulted virtually every possible political, racial, or religious camp of the day. That said, he wrote about an important subject at a time when written works could perhaps make a difference. While the world he lived in was swooning at the technological marvels of the day, he wrote this about the philosophical culture and the era immediately preceding his own: "In the place of the authentic religion of earlier times came a shallow enthusiasm for the 'achievement of humanity', by which nothing more was meant than progress in the technics of labour-saving and amusement-making. Of the soul, not one word was discussed."[137]

Here was a man ready to stop wondering at the sparkles, spangles, and blinking lights of technology long enough to ask, "What's really happening to our souls in the midst of all this?" He provided such a promising beginning to his book, and he even followed that auspicious opening with many accurate predictions about where we would go collectively with technology, predictions that today seem almost de Tocqueville-like in their accuracy.

Not to get ahead of ourselves, but we'd like to note one very important difference between de Tocqueville's predictions and those of Spengler: de Tocqueville's tone was one of humility. He said at one point in the Preface to *Democracy in America*, "I am ignorant of his [God's] designs, but I shall not cease to believe in them because I cannot fathom them, and I had rather mistrust my own capacity than his justice."[138]

We'll come back to that attitude of humility—and its role in our relationship to technology—after a review of Oswald's own sometimes amazing insights.

Spengler passed away in the early shadows of WWII, in 1936. He had not yet seen the atom bomb, which was about a decade away, nor had he seen supersonic flight, space exploration, or ENIAC.[139] He had the insight, however, to understand that many men of his era were making decisions about technology from, at best, an awkward philosophical footing. He noted the shallowness of the stance toward technology that belonged to some of his philosophical predecessors in Bentham, Mill, and Spencer, attributing the following concept (and impacts on the attitudes of society) to them: "The aim of mankind was held to consist in relieving the individual of as much of the work as possible and putting the burden on the Machine. … The progress-philistine became excited over every button that set an apparatus in motion for the—supposed—sparing of human labour. … In the place of the authentic religion of earlier times came a shallow enthusiasm for the 'achievements of humanity'.…"

"Authentic religion" lost and an alleged "shallow enthusiasm" spread across an entire era? Both are heady assertions. Yet with hindsight, we look at the headlines and noted feats of men in the 20th century and notice they do, unmistakably, revolve around technology. If one of Spengler's main points is that men in the 20th century had a skewed understanding of the role they should grant technology in their lives, we could hardly dispute his point. For Spengler (as perhaps for Ellul and his *la technique*), those of us born in the 20th century need a completely new understanding of technology:

> If we are to understand the essence of technics, we must not start from the technics of the Machine age, and still less from the misleading notion that the fashioning of machines and tools is the *goal* of technics. ...the significance of technics may only be seen in terms of the *soul*. ... *Technics is the tactics of all life.* It is the inner form of the *process* utilised in that *struggle which is identical with life itself.* ... *Technics is not to be understood in terms of tools.*[140] [Emphasis in the original.]

Was Spengler merely concerned about people mistaking "tool making" as the goal of technology? Actually, his argument was much deeper than that, and in some regards, Spengler makes the very point that forms the backbone of our thesis here: understanding why and how we use any given technology is important because we *use technology as an extension of ourselves.* Spengler's seemingly extravagant talk of "the soul" in relation to technology is not off base at all when viewed in the light of our original and genuine goals in developing technology. He brings us back to the very human roots of technology with statements such as: "Every machine *serves* some one process and owes its existence to *thought about this process.*"[141] [Emphasis in the original.]

And that last point is where Spengler's regard for Nietzsche's thought comes into play. Spengler addresses previous generations' carelessness (and/or misunderstanding) about our use of technology, saying, "It was a little ridiculous, this march into the endless future, towards a goal which men did not seriously conceive or dare to visualise clearly. For by definition a goal is an end. No one does anything without thinking of the moment when he shall have attained that which he willed."[142] Having touched on Nietzsche's core theme, *will to power*, Spengler uses the rest of *Man and Technics* to convince readers how critically important will to power is. He even adopts a blatantly materialistic stance (in apparent contradiction to his earlier concerns about disregard for the human soul), saying "Property is the domain in which one exercises unlimited power.... It is not a right to mere possession, but the sovereign right to do as one wills with one's own."[143]

As Spengler builds toward his final prediction of the demise of "the machine culture," as he calls our current technology-centric era, he takes the will to power to a seemingly logical, and in his thinking, *inevitable* conclusion:

> But it is here, in our own [millennia], that the struggle between Nature

and the Man whose historic destiny has made him pit himself against her is to all intents and purposes ended. ... A will to power which laughs at all bounds of time and space, which indeed regards infinity as its specific target, subjects whole continents to itself, eventually embraces the world in the network of its forms of communication and intercourse, and *transforms* it by the force of its practical energy and the gigantic power of its technical processes.[144]

Finally, Spengler builds to the blasphemous end of the will to power, and specifically the will to power expressed through technology:

To build a world *oneself*, to be *oneself* God—that is the Faustian inventor's dream.... Finally, with the coming of rationalism, the belief in technics almost becomes a materialistic religion. Technics is eternal and immortal like God the Father, it delivers mankind like God the Son, and it illumines us like God the Holy Ghost. And its worshipper is the progress-philistine of the modern age....[145]

He admits, however, there is a price to pay for blasphemy: "This machine technology will end with the Faustian civilization and will one day lie shattered and *forgotten*.... When, and in what fashion, we know not."[146] [Emphasis in the original.]

Along the way, Spengler mentions the usual figures quoted in discussions about man, technology, and society—Nietzsche, Bacon, Faust, etc. On the very last page of his book, having embraced various elements of their thought and supposed insights, he concludes, almost like a Nazi soldier in 1945 with his back against a bunker wall: "Only dreamers believe in ways out. Optimism is *cowardice*. We are born in this time and must bravely follow the path to the destined end. There is no other way. Our duty is to hold on to the lost position, without hope, without rescue."[147]

"The destined end?" Let's return for a moment to the issue of humility. We quoted at the outset of this chapter de Tocqueville's own clear statement of faith in someone greater than man or history. More recently, Popper indicated in his philosophical works a similar, albeit less direct belief that man is *not* the director of history; by embracing only *piecemeal* social engineering and rejecting the more familiar and nearly universally invoked utopian social engineering, Popper is saying, among other things, that he does not believe man can rationalize his way to heaven on earth. While his distinction between the two forms of social engineering may not be a statement of faith in God, per se, it is an increasingly rare admission by a man of letters that Socrates' epistemic humility is still alive and well in some corners. Popper, perhaps, is saying that men can form, live, and work within social institutions without believing that we will become genuine and comprehensive masters of the world *through* those institutions.

Spengler and de Tocqueville both describe their vision of the future of society. Neither man is particularly optimistic about that future. Again, though, we see an important difference in Spengler's tone. His pronouncements about the future are unapologetically deterministic in character.

Popper, for one, expressed his concerns about determinism. He openly questioned those who claim to have found inviolable laws within history, and who, in the name of those laws make prophet-like predictions about the future. While Popper does not (sadly) devote whole chapters to discussing the "crutch" of reification, Spengler and many other writers use that crutch to the detriment of their arguments. Does it really matter, one might ask fairly, if Spengler—or any other writer—mistakenly attributes human-like will and volition to concepts such as history or technology? This question is perhaps all the more salient given how correct Spengler seems to have been with so many of his deterministic predictions.

One partial answer to that question might be that, yes, it certainly matters, because Bernays and many others after him have shown that when the masses adopt a view of *inevitability* about certain technological trends (*"everyone's doing it..."*), they are more likely to adopt Spengler's "all is lost" bunker mentality when faced with new technologies, even when faced with frightful technologies such as nuclear-tipped missiles. That is to say, those who accept technological determinism as their default mindset are not very likely at all to take philosophical, conceptual, or moral stands about new technologies. Why should they? It's coming anyway, right?[148]

There is also, perhaps, a self-fulfilling aspect to making god-like pronouncements about technology's future direction. Worst of all, once a society adopts a deterministic stance, the philosophical, era-spanning pronouncements of a deeply thinking and broadly influential Spengler, Nietzsche, or Marx no longer seem necessary for keeping that mindset intact.

A few examples of the "alive and well" tech-determinism mindset follow. We'll start with the titles of a few articles culled from the last few years of (mostly) online journalism:

- "Easy DNA Editing Will Remake the World. Buckle Up."

- "McCain: Americans Should 'Accept' That Their Private Conversations Are Being Monitored"
- "US Company Gives Glimpse into Future of Government Surveillance"
- "Drone Revolution Hovers on the Horizon"
- "Transhumanist Kurzweil Predicts Human/Computer Hybrids by 2030s"

Note the "inevitable" tone of those headlines. "Buckle up." "Revolution Hovers."

And now some select quotations:

On Cloud computing: "Somebody is saying this is inevitable – and whenever you hear somebody saying that, it's very likely to be a set of businesses campaigning to make it true."[149]

On transhumanism: "Kurzweil...[said] that people will soon be able to connect their brains directly to the cloud, and augment their existing intelligence with thousands of computers. 'Our thinking then will be a hybrid of biological and non-biological thinking,' Kurzweil said, explaining that the brain will use DNA nanobots to make the connection. 'We're going to gradually merge and enhance ourselves'"[150]

On artificial intelligence: "[Bill] Gates went on to say that the profound impact from this technology is set to accelerate over the next ten years and make it 'strong enough to be a concern.'"[151]

I once listened to a "futurist" at a tech conference announce at the beginning of his talk that he was "retiring." He was billed in the convention agenda as a futurist, and he gladly accepted the moniker during his introduction. So, how *does* a futurist retire? Is that a bit akin to a god announcing his retirement? I remember wondering if the futurist's retirement meant the future was now going to change?

Snide comments aside, we do have to wonder at ourselves for accepting the very concept of tech "futurists" in our midst. If we read in our history books that, say, during the post-colonial period of the US, men went around pronouncing the future to large crowds, and furthermore, that people in those crowds went about making massively impactful decisions about how they would communicate, how they would run huge portions of their material lives, and even how they would structure both their business *and* personal relationships based on what these would-be prophets said, we'd wonder at the sanity of the era.[152] And yet, technological futurism is the fad of our day. (At least we can hope it's only a fad.) Who could have known that the sci-fi genre was really just a form of newsletter? Jules Verne = Johannes Gutenberg, more or less.

Hopefully at this point we've come full circle to digital reliability. How so? Recall that we said that in measuring whether digital technology is reliable, we believe that we should "determine where digital technology is inducing indifference, inactivity or incapacity where human action should be." We can rightly ask if we're denying ourselves the opportunity for right action when we allow others to pontificate about our technology future without responding. In many cases, it seems we don't even show adequate *skepticism* in the face of "inevitable" movements such as genetically modified food, animals, pathogens, and humans, or the filling of our highways with driverless cars and our skies with drones.

Simplistic though this may be, Nietzsche, and by extension, of course, Spengler, would have us say that since "God is dead," our own will to power remains the arbiter of goals and ends. Do the "people," the masses, exercise that will to power? Or how about an Ubermensch? Do we subsequently make our way through the future via "futurists"? Have they replaced the Oracles, the Prophets, and even that poor human substitute, rationality itself? If so, we certainly need a hiatus in our "march forward" so that we can at least grant ourselves a chance to pick through the psychological, religious, and social detritus, to catch our collective breath, to embrace this self-making of humans as gods. Wow. To think we only needed wait for a Nietzsche or a Spengler to grant us technical/secular powers and rights with Zeus and Prometheus, or, if you prefer, with St. Peter and St. Paul. When will we transfer the keys to the Kingdom from St. Peter to a futurist?

We Have a Choice

Every day, most of us have some choice about what we do: waking up, brushing our teeth, going to work, and choosing to go for a jog (or not). If we stop and ask ourselves why we do those things, we usually have answers: If I don't get out of bed now, I'll be late for work; If I don't brush my teeth, I'll have painful issues in the future; If I skip my jog consistently, then I may need to save up to buy some bigger pants. Similarly, every piece of technology someone developed through a series of human choices, beginning with a set of tasks or problems the technology aims to address— in a sense its reason for being.

108

In this book, we don't pretend to predict the future, and we understand that perhaps 20 or 30 years from now, reading our concerns in these pages will bring on a good laugh. That does not change the fact that every generation faces big questions about right and wrong, about what we're doing that impacts ourselves, our society, and even our survival.

We seem to be on the verge of living out some of our most popular science fiction ideas, so it's understandable that many of us feel almost giddy about new digital technologies. In the midst of our joys with new toys, we should keep in mind that these new technologies force us to make many choices about how we spend our time. And these choices do not always produce clear-cut outcomes. For example, brushing our teeth means we can chew food, but what do we *really* understand about what using Facebook three hours a day really does for (or *to*) someone?

The overall gains versus costs of switching from older to newer technologies are sometimes less than obvious and, of course, may be viewed differently by different people. For instance, I find that reading a physical book only for enjoyment is a much more personal and singular activity than reading a website online or a Kindle-based book. For some, the relative isolation of reading a printed book is a wonderful side effect of yesterday's technology (i.e., the printed word). For others, reading in isolation means needlessly removing opportunities for sharing instant feedback and digital conversations about the subject at hand.

Yet, reading everything in digital format has invisible costs. Yes, online reading is more convenient, arguably more portable, and certainly offers more opportunities for real-time social interaction than does reading the traditionally printed book. (And, yes, the same technology allows people like us to self-publish.) So, there's the good. The *bad* is that we are crossing a threshold in which almost everything we do now in the digital world comes at the cost of exposing personal data that was quite private in pre-digital times: your communications, your habits, your interests, and even your location. The many printed newspapers and magazines I've held in my hands did not report back to a server how long I lingered on a page nor how many ads I glanced at while reading.

Most today will shrug at this notion of invisible cost. We believe it's time to stop shrugging. If the thinkers of the past or the dystopian visions of some futurists don't scare you, then we recommend looking no further than current events. Pay attention to the daily articles related to automation failures, social media blunders, banking system crashes, privacy breaches, and governments spying on their own. Recognize that tech-determinism is too often an excuse and is never an absolute—that we have a choice about these technologies and how they are developed and accepted into society (or not). It's "OK" to demand change in the way we make decisions about technological development.

The good news is that over the last few years, while collaborating on this effort, we have seen an increase in the number and visibility of people questioning where technology is headed (many of whom we quote in this book). Bill Gates, Elon Musk, and Stephen Hawking, for example, have each warned about the potential dangers of artificial intelligence. Yet despite the warnings of these well-known technologists and B-list celebrities, we see an unrelenting push towards total digitization, and this push comes without the slightest sense that we still have a choice in the matter.

The same global and digitally connected technologies we celebrate for their ability to illumine those behind various latter-day Iron Curtains could easily undermine the societies who shrug at or do not understand the potential for misuse of those technologies. Considering changes in technology more holistically, discussing potential benefits while also evaluating costs, understanding the "whys," and mitigating the potential for misuse—all of these are vital starting points if we are to enjoy the fruits of well thought out and responsible technological deployment.

We find it easier to fear a leaking nuclear plant or an act of terror involving a dirty bomb—both events that are highly unlikely to impact any given individual, statistically speaking—than we do the impact of placing cellphones in the hands of our children. Yet we're accumulating more and more evidence that digital dementia and generally fragmented thinking and concentration among young people is a direct result of excessive use of digital technology. And the revelations of government insiders such as Binney and Snowden show that governments willingly turn digital technology on their own citizens. Perhaps our era needs an updated version of James Madison's saying, "The means of defense against foreign danger historically have become instruments of tyranny at home."[153] So often, we in our era fear terrorism from abroad while we steadfastly ignore types of technology-based tyranny at home.

No Fat Lady in Sight

When we originally set out to write this book, we planned to highlight several well-known (and some lesser known) disasters caused by digital technology failures, along with the unanticipated consequences of those failures. We've shared a number of disaster examples, but sadly, we have so many more to review.

We came together as co-authors in part because we worked together in software testing, putting enterprise-level systems through the paces for some of the largest DoD clients in the world. We had in common a first-hand understanding of just how fragile and fickle digital technology can be.

We *did not* have in common religious, political, or philosophical backgrounds, nor do we come from the same generation. As such, we do not always agree on every issue. Yet, these acknowledged differences in background and outlook augment, to some extent, the success of one of our goals in writing collaboratively; we'd very much like to point out that the influence of digital technology, as we are implementing and integrating it today, cuts across generations, politics, gender, and generation. If, as the tech corporations love to insist, they are building and releasing "life-changing" devices and systems, then all of us, regardless of background or ideology, should have our eyes and ears open for those aspects of our lives that are indeed changing.

One of our main goals is to jolt those of us who work in science, technology, or engineering to see a bigger picture regarding potential consequences of failed or misused technology. In the years we've worked on this book, we have heard more and more voices join the chorus related to technology and its impact on our privacy, safety, quality of life, and our freedoms. So many stories hammer home similar cautionary tales that, in fact, we find that it is increasingly difficult to stay on top of them all. We still have a lot to say, and we plan to say it in additional parts to this book.

After much discussion, however, we decided that we can no longer wait to publish what we have on these pages. We still have our day jobs, and given the normal time demands of job and family, we decided to leverage digital publishing (the irony is not lost on us) to distribute the first version of this book. The pace of technological change appears to be increasing, so we choose not to wait any longer to publish. Even if you're reading this in print format, we had to use digital processes to make it happen!

We are not trying to stop the world from changing. But, we believe it is necessary to change our world more thoughtfully.

Please visit **www.MIAATQ.com** (or www.Cyberskeptics.net) to start a conversation with us, to join our efforts to spread the word, and to watch for the release of *MIAATQ: Volume II*.

APPENDIX A: KEY DEFINITIONS

Editor's note: For the most part, the key definitions in this appendix comprise the authors' understanding of these terms, particularly as we use them in MIAATQ. The definitions themselves are micro-essays, and we intend to publish them on MIAATQ.com, likely in an even further expanded format, for the benefit of exploring these subjects in greater depth.

Ambiguation of Roles

Derived from "If You See Something, Say Something," in "Surveillance and Society," 10(3/4): 235-248, by Joshua Reeves:

The process of confusing traditional roles within a given society; the leveraging of social role versatility for intentional purposes of behavioral manipulation.

Technology has been impacting and changing the roles of human workers at least since the first fire was used to warm a cave or a tent. The human confusion that sometimes results from these changes we call "ambiguation." Sometimes ambiguation is part of societal violence (as it was in the Luddite controversies of 19th century England). Other times we simply change our understanding of our jobs and roles in our families, communities, and professions when faced with technology-induced forms of ambiguation. In any case, the impact of ambiguation will likely increase in the near term as scientists and technologists move from simply using machines and technology in order to augment physical activity (e.g., digging a hole, lifting heavy objects, traveling at high speeds) to the goals of emulating human mental, communicative, and affective activities via computers, machines, and robots.

In "May I Ask a Technical Question," we take an in-depth look at ambiguation because we believe that when it arises from attempts to emulate the human mind, ambiguation almost always contributes to the mechanization of the [human] mind. And this process is so subtle, we detect it rarely, at best.

In short, ambiguation in a world full of technicians striving to create artificial intelligence has the potential to trigger fundamental changes in our society and, ultimately, in our understanding of ourselves.

Artificial Intelligence

From http://www.merriam-webster.com/:

An area of computer science that deals with giving machines the ability to seem like they have human intelligence.

The field of Artificial Intelligence (AI) seems to engender few neutral reactions. The concept, if not its fruition, is simple enough: build a machine, or system of machines, that can mimic human intelligence; "Alan Turing's 1950 article "Computing Machinery and Intelligence" discussed conditions for considering a machine to be intelligent. He argued that if the machine could successfully pretend to be human to a knowledgeable observer, then you certainly should consider it intelligent." From http://www-formal.stanford.edu/jmc/whatisai/node1.html [Editor's ironic note: I've been pretending to be human for much of my adult life...]

From its simplistic beginnings, the ~ 60-year history of the quest for AI devolved rapidly into confusion, at best, and combat-like proclamations, at worst. Langdon Winner notes that lovers of the pre-determined tech-utopian future (here's looking at you, Silver, Stock, Kurzweil, Moravec, and Warwick) see inevitable progress to and for AI-infused Metaman, while equally venerable AI pioneers such as Joseph Weizenbaum say the idea of making a non-biological machine intelligent is "obscene, anti-human and immoral." (See Stanford.edu reference in the previous paragraph.)

For our purposes in MIAATQ, we emphasize the word "seem" as it appears in the definition of AI found at http://www.merriam-webster.com/. (We quote m-w.com for our simple definition of AI in the Table of Concepts, found at the front of this book.)

Here, then, is something of a conundrum: Human intelligence may, apparently, cease to be either human or intelligent when once defined. Yet, without a starting point that includes the definition of intelligence, how can logic-based systems begin to replicate it? What is left, it seems, is to seem.

Automatons

From http://www.merriam-webster.com/:

1. A machine that can move by itself.
2. A person who acts in a mechanical or machinelike way.

Automata, as man-made curiosities, have been around for many hundreds of years. They are human-made objects intended to mimic certain human movements and abilities; yet, through the years, many creators of automatons have decided—in violation of the informal rules about building automata—to augment their abilities surreptitiously.

(See https://en.wikipedia.org/wiki/The_Turk for a famous example of duping the public about the real capabilities of a particularly famous "chess-playing" automaton.) We see an irony in stating one goal for creating automata (e.g., precise human simulation without run-time human intervention), while quietly undoing that goal in the background. We revisit salient facets of this irony below, as well as in several chapters of MIAATQ. For our purposes in this book, we're not as concerned about the detailed differences between automata and robots. The important aspects of our definition of automatons touch more on debatable points of view, such as that of Professor Hiroshi Ishiguro, himself a developer of advanced robots. He recently was quoted as saying, "Robots are a mirror for better understanding ourselves. We see humanlike qualities in robots and start to think about the true nature of the human heart, about desire, consciousness and intention."

(See http://www.theguardian.com/technology/2015/dec/31/erica-the-most-beautiful-and-intelligent-android-ever-leads-japans-robot-revolution)

Professor Ishiguro, as a *creator and designer* of robots with a stated goal of trying to create robots that resemble humans even in the details of how they interact with us, must certainly have a need to reflect on the nature of the human heart and consciousness. It's fair to ask, however, if the millions of consumers expected to buy personal robots in the near future will be inspired to do any of the same sort of deep reflection about how they use and interact with their robots.

Our concern is that as we engage in unconsidered employment of robots in large areas of society, we might realize this frightening portion of Merriam-Webster.com's simple definition of automatons: *a person who acts in a mechanical or machinelike way*. Stephen Talbott expressed the real depth of this concern when he said, "Not only did the machine originate with us, and not only does it live in us, but now it has a massive external and objective presence in our lives. We can interact with all this machinery only by shaping ourselves to its requirements. ... The danger of self-forgetfulness, then, is the danger that we will descend to the level of the computational devices we have engineered—not merely imagining ever new and more sophisticated automatons, but reducing ourselves to automatons." This potential tragedy may reflect the kernel of the irony mentioned above, i.e., that creators of automata seem to yield to a consistent temptation to provide unseen human intervention into the run-time behaviors and operation of the automatons. And this temptation may signal the far more dangerous trap of shaping ourselves to our machines, rather than the reverse.

Control Centrism

From the authors of MIAATQ:

The belief that society and living environments can and should be controlled via centralized decision-making akin to that seen in most models of digital technology.

The term *control centrism* appears at least twice in the book *Modern Governance – New Government-Society Interactions*, edited by Jan Kooiman (Sage Publications, 1993). One of the book's featured authors also references a previous publication, circa 1983, that refers to "control-centrism" as the "tendency to regard government as the axis around which the world turns..." (From an essay by Marijke Prins, "Women's Emancipation as a Question of Governance: Actors, Institutions and the Room for Manoeuvre," referencing Den Hoed et al., 1983: 43).

The pre-Internet references above notwithstanding, MIAATQ identifies Control Centrism as evolving lock-step with the expanding use of digital technology throughout the world. Digital technology, by definition, *quantifies* the world. And in the 20th century and the early decades of the 21st century, this quantification is joined by the de facto rejection of mysticism, revelation, metaphysics, or any belief system showing significant deviation from the tenets of logical positivism. The authors of MIAATQ deem this one-two punch of quantification and verificationism to be a daunting challenge to the many tenets of philosophy, theology, and anthropology that venerable thinkers and writers throughout the ages have put forth in the hope of keeping humans free to learn and to grow in many *un-*quantifiable and non-political ways.

Cybernetics

From http://www.merriam-webster.com/:

The science of communication and control theory that is concerned especially with the comparative study of automatic control systems (as the nervous system and brain and mechanical-electrical communication systems).

116

Jean-Pierre Dupuy, accessed from:
http://www.metanexus.net/print/essay/h-cybernetics-antihumanism-advanced-technologies-and-rebellion-against-human-condition, 19 December 2015, states:

Cybernetics calls to mind a series of familiar images that turn out on closer inspection to be highly doubtful. As the etymology of the word suggests, cybernetics is meant to signify control, mastery, governance—in short, the philosophical project associated with Descartes, who assigned mankind the mission of exercising dominion over the world, and over mankind itself. Within the cybernetics movement, this view was championed by Norbert Wiener—unsurprisingly, perhaps, since it was Wiener who gave it its name.

For our purposes in MIAATQ, our caution related to cybernetics is summarized in this further quote from Dupuy:

The philosophers of consciousness were not alone in being caught up in the trap set by a question such as 'Will it be possible one day to design a machine that thinks?' The cybernetician's answer, rather in the spirit of Moliére, was: 'Madame, you pride yourself so on thinking. And yet, you are only a machine!' The aim of cognitive science always was—and still is today—the mechanization of the mind, not the humanization of the machine.

Dupuy uses somewhat ironic wording here—cognitive *scientists* generate *aims*, and these aims are not generated by the science itself; we should exercise caution to avoid implying sentience in a method and, thereby, denying the men and women using that method the responsibility for how they use it.

Yet, Dupuy's point aligns with one of our major concerns in MIAATQ: we have to remain ever-vigilant that we understand both our (genuine) aims and (actual) consequences within our work with digital technology. More than one author expresses concern that as we "digitize thinking" within machines, we are mechanizing ourselves. As Stephen L. Talbott notes—without, in our opinion, the least bit of hyperbole—digitizing thinking places us in danger of becoming automatons.

Determinism

From http://www.merriam-webster.com/:

1. A theory or doctrine that acts of the will, occurrences in nature, or social or psychological phenomena are causally determined by preceding events or natural laws.
2. A belief in predestination.

Determinism, for our purposes in MIAATQ, refers to a strand of belief running through a large set of ideas and philosophies that human events and movements unfold with the help of an unseen and impersonal hand. In common modern usage, those invoking determinism usually omit reference to God's hand or direct divine intervention, and this may be a reflection of the general growth of secularism in the West in the last several hundred years. Thus, absent the more traditional belief in an intervening God or gods, many of our encounters today with determinism in philosophy, literature, and even in theology are related to reification, in which people use language that seems to grant life, abilities, will and/or sentience to abstractions or objects. (Note: The less consciously aware we are of our encounters with determinism and reification, the more likely we are to compound human mistakes related to *not* taking proper responsibility for our actions and choices. Explanation follows.)

Two clear examples of determinism in modern thinking can be seen in:

1) Marxist ideology ("history," *sans* God or gods, as a constantly moving force with a teleology).
2) Hegel's writings (epochs of human history containing their own logic).

In both of these instances, those who believe they observe a quasi-sentient "history" or "epoch" provide selected opportunities to absolve themselves and others of the *human* responsibility to attempt social or political change; or worse, they grant themselves a sort of socially dangerous carte blanche— the most literal example of conducting oneself as if the "ends justify the means," i.e., those who believe they understand the teleology of a society (or history, or an epoch...) by definition believe they understand *its ends*. Once the crucial projections about teleology have been made, it's only a short step to take up Marx's task of lessening the "'birth pangs' of future social and political developments." This mentality and activity, i.e., pushing the events and shape of society in what one believes to be a predetermined direction, "leads inevitably to totalitarianism and authoritarianism—to centralised governmental control of the individual and the attempted imposition of large-scale social planning" (From *Stanford Encyclopedia of*

Philosophy, "Karl Popper," accessed from:
http://plato.stanford.edu/entries/popper/#SocPolThoCriHisHol, 12
December 2015).

In short, many of today's tech thinkers and writers embrace forms of
determinism openly, and in doing so, some believe they assert that
technological "progress" (like Marx's concept of history before it) is
inevitable and unstoppable—via human will alone. This belief is sometimes
implied, as when consumers, without pause, continually seek out the latest
electronic gadget; at other times, "futurist" writers such as Kurzweil,
Moravec, or some of the transhumanists very explicitly spell out their belief
in unstoppable technological progress, with Gregory Stock's *Metaman: The
Merging of Humans and Machines into a Global Superorganism* providing a perfect
example of this type of thinking.

Digital Dementia

From https://www.psychologytoday.com/blog/mind-
change/201507/digital-dementia:

*Digital Dementia is a term coined by neuroscientist Manfred Spitzer to describe an
overuse of digital technology resulting in the breakdown of cognitive abilities [Referencing:
Spitzer M. (2012). Digitale demenz. München: Droemer, 7].*

The simple definition of dementia found at Merriam-Webster.com/ is:
"*medical*: a mental illness that causes someone to be unable to think clearly
or to understand what is real and what is not real."
While the term "digital dementia" is not yet found in many of today's
lexicons and dictionaries, a growing number of studies show an increasingly
clear connection between some forms of dementia and excessive use of
digital technology.

Early studies of this affliction, such as those conducted on populations of
users of digital technology in South Korea, relied mostly on tests of
memory and measuring declines in human cognitive ability. A recent article
in "Psychology Today," however (see the URL above), references several
studies that use direct research on changes in brain volume sizes triggered
by use of digital technology. Thus, while many remain skeptical about how
genuine is this newly identified affliction, both epidemiological studies and
investigations of physiological changes seem to confirm that overuse of
digital technology induces mental impairment that would otherwise not be
present.

Digital Reliability

From the authors of MIAATQ:

The degree to which we can trust that digital technologies will work reliably and neither operate nor break down in ways that jeopardize our individual and societal health, sanity, safety, and comfort, nor function in ways that enslave us to any degree.

In some ways, defining Digital Reliability is the point of MIAATQ. The authors believe we should measure the digital wonders showing up continually in our lives not simply by their surface abilities and the individual tasks they perform for us, but also, just as importantly—every bit as necessarily—we should measure and consider what human and social tasks, abilities, traditions, skillsets, institutions, and opportunities digital technologies and devices displace and impact.

Thus, the authors' expanded definition of Digital Reliability includes:

1.   Considering, on an ongoing basis, whether a given digital technology, taken as a whole and viewed in the context in which it operates, is hurting ourselves or others.
2.   Measuring and documenting, to the extent possible, where digital technology is failing us in previously unnoticed, unobserved ways, *whether or not* those failures relate directly to the intended functionality designed into the digital technology.
3.   Determining where digital technology is inducing indifference, inactivity, or incapacity where human action should be.

Digital Technology

From http://whatis.techtarget.com/definition/digital and the authors of MIAATQ:

Digital technology is electronic technology that 1) generates, stores, and processes data in terms of two states: positive and non-positive and that 2) depends upon quantification of all input in order to operate.

We mean to use the term "digital technology" in MIAATQ in a very broad sense. On the one hand, as we write this book, computers—more specifically personal computers—represent the most ubiquitous and familiar form of digital technology in the world. On the other hand, that may change soon. Multiple hardware vendors are trumpeting the "Internet of Things (IoT)" and other concepts and products aimed at inducing, in addition to sales, broad public acceptance of digital technology in the products and environments of everyday living.

Norbert Wiener, and later Neil Postman, identified computers as the technology of "command and control." While computers remain just that, whether employed in eyeglasses, wristwatches, self-driving cars, or "smart electric meters" [sic], both government and industry in the US continue to push for broad acceptance of digital technology as *the way* for humans to communicate and relate to one another.

If the well-publicized warnings of Bill Gates, Elon Musk, Eric Schmidt, and Stephen Hawking are correct, our collective work with artificial intelligence may soon serve to expand further the definition of digital technology. AI, according to these well-informed gentlemen, will soon leapfrog our comfortably accepted consumer- and social-oriented computers to become the most familiar—and feared—form of digital technology.

Gnomic Will

From http://usqr.utsnyc.edu/wp-content/uploads/2012/03/Blowers.pdf:

Modern lexicons confirm the word's pliability, being variously translated "mind," "will," "purpose," "intention," "inclination," "opinion," "character," and more.

[Editor's note: Apologies for using a term and definition from theology here. The concept of gnomic will is not necessarily new, as the reference to Aristotle below confirms. Theology, in this case, however, provides us with the clearest definition for purposes of how gnomic will is salient to our discussions about digital technology.]

We differentiate between the "natural will" and the "gnomic will," the gnomic will being far more familiar to us, once we define the term; gnomic will involves our use of deliberation, rationalizing, and calculating in order to make choices and decisions about any given path or activity in life. The natural will—a term recognized currently by only a relatively small subset of Christians—by contrast, is the type of will that an omniscient God exercises, knowing at all times which path to take, and thereby having no need for rationalization or internal deliberation. The following quotation is from: https://en.wikipedia.org/wiki/Gnomic_will, accessed 1 November 2015:

The notion of gnomic will belongs to Eastern Orthodox ascetical theology, being developed particularly within the theology of St Maximus the Confessor. The term 'gnomic' derives from the Greek gnome, meaning 'inclination' or 'intention'. Within Orthodox theology, gnomic willing is contrasted with natural willing. Natural willing designates the movement of a creature in accordance with the principle (logos) of its nature towards the fulfillment (telos, stasis) of its being. Gnomic willing, on the other hand, designates that form of willing in which a person engages in a process of deliberation culminating in a decision. ...

According to [the theology of] St Maximus, the process of gnomic willing presupposes that a person does not know what they want, and so must deliberate and choose between a range of choices. However, Jesus Christ, as the Second Person of the Holy Trinity was omniscient. Therefore, St Maximus reasoned, Christ was never in a state of ignorance regarding what he wanted, and so never engaged in gnomic willing. Aristotle, in comparing the works of Nature with those of a human worker, had also declared that any process of deliberation, far from indicating superior intellect, is a sign of our weakness.

Why do we include the concept of *gnomic will* in a book about digital reliability? Glad you asked. Again, within MIAATQ, one of our main goals is to look for unintended consequences of "advances" in digital technology. What if human beings *are* sometimes capable of making good decisions without "normal" deliberation, perhaps via revelation or some form of spiritual guidance—a human capability in which millions upon millions, of virtually all religions, have believed throughout the ages.

If we can sometimes "know" the correct path in life without employing syllogisms and various forms of logic and deliberation, then our *ipso facto* designation of computer logic as the "way to think" may induce us to ignore or reject an entire range of our human capabilities and deeply important elements of our nature. If digital technology "institutionalizes" logic at the expense of all other forms of human knowing and willing, if human decision-making increasingly comes down to weighing different paths and options in terms of "1s" and "0s," then *gnomic will* versus *natural will* may be the single most important concept to consider as we build digital technology.

Joseph Weizenbaum, a world-class pioneer (and eventual critic) of artificial intelligence, may not have heard of *gnomic will*, per se, but he noted—very much in line with the Christian contrast between gnomic will and natural will—the crucial distinction between deciding and choosing: "Deciding is a computational activity, something that can ultimately be programmed.

Choice, however, is the product of judgment, not calculation." (From
https://en.wikipedia.org/wiki/Joseph_Weizenbaum)

Hegelianism

From http://www.merriam-webster.com/:

*The philosophy of Hegel that places ultimate reality in ideas rather than in things and
that uses dialectic to comprehend an absolute idea behind phenomena.*

Reviewers of MIAATQ noted that we take some liberties in the book in our
discussions of Hegelianism. Our best rejoinder may be to note that we
believe we're reflecting accurately the very liberties that some leaders exploit
when using the Hegelian dialectic for purposes of social engineering.
As Karl Popper and others note, Hegel's writing is nearly opaque, even for
academics. Nonetheless, many have latched onto his concepts of dialectic
and triadic development. Whether or not academics (or we) understand
Hegel's original intent, we can observe use of a form of triadic development
almost daily in the political realm, however impure it may be compared to
Hegel's original concept.

We believe a variation (or watered-down?) version of triadic development is
particularly visible today when we consider the nearly ubiquitous
combination of digital technology and planned obsolescence, along with the
inherent motivation of many in modern industry and government to
simplify, and perhaps to hasten, implementation of centralized control via
digital technology.

Historicism

From http://www.merriam-webster.com/:

1. A theory, doctrine, or style that emphasizes the importance of history.
2. A theory in which history is seen as a standard of value or as a determinant of events.

Historicism is often confused with the word "historism." In MIAATQ, we
use the term "historicism" in accordance with Karl Popper's understanding
and definition of the word, and we do not use the word "historism."
Historicism is, from one point of view, a variety of determinism.
Determinism adopts the singularly secular point of view that trends and

movements in society and culture come about through the influence of "laws" or "principles," seen or unseen; and these laws and principles have nothing to do with the personal nature—the "personhood," if you will—normally attributed to God (or gods). For historicists, "history" embodies impersonal laws and principles that govern life.

Historicists, then, grant sentience and the ability to influence the events and lives of humans and their societies to something that was, traditionally, simply a concept—namely, history. As Popper notes, with this concession in place, the achievement of uncovering and understanding the underlying laws and principles of history grants their potential "discoverer," ironically, almost god-like abilities to predict and prescribe the events and trends of his fellows. The irony comes from the fact that many historicists begin by embracing a decidedly secular world view that excludes "gods" and God, and they end up embracing a conceptual framework that confers upon them Oracle-like abilities: "These [historicist] beliefs lead to what Popper calls 'The Historicist Doctrine of the Social Sciences', the views (a) that the principal task of the social sciences is to make predictions about the social and political development of man, and (b) that the task of politics, once the key predictions have been made, is, in Marx's words, to lessen the 'birth pangs' of future social and political developments." (From Stanford Encyclopedia of Philosophy, "Karl Popper," accessed from: http://plato.stanford.edu/entries/popper/#SocPolThoCriHisHol, accessed 12 December 2015).

We include historicism as a topic of importance in this book because so many writers, observers, and pundits in technical journals and literature combine reification, determinism, historicism, and solutionism into a lethal blend of fantasy in which human responsibility for action and self-control can be shuffled aside in a sort of non-stop shell game; so long as there is another "force" to look to for the direction of the future, we need not make difficult decisions nor even give extended consideration to what we're willing and building today.

Incrementalism

From http://www.merriam-webster.com/:

A policy or advocacy of a policy of political or social change by degrees.

We deem incrementalism an important concept in the study of digital reliability because those who employ it often seem to do so with agendas in mind that go far beyond simply providing a solution to a human problem or aiding in the completion of a set of tasks. And aren't problem solving and task completion supposed to be the relatively simple goals of digital technology? Insofar as agendas displace human problem solving and task completion, we can see the soil being prepared for unannounced consequences.

For example, if one of the goals of a major software vendor is to create a sustainable market share and to generate oft-returning customers, then the vendor may have little incentive to create a "perfect" word processor. That is to say, a word processor that is feature rich and solves many problems for writers and does so consistently and with little or no maintenance—such a product may not solve the *vendor's* business problem at all. Customers who are given a stable, dependable, and usable tool may not return to the original vendor as often as the vendor would prefer (for purposes of sustaining cash flow, market share, and other typical goals of modern corporations).

Technopedia.com, in its definition of the software term "bloatware," noted the following fallout when tech companies employ incrementalism: 'Bloatware usually occurs as a result of feature creep. Because software is traditionally redesigned on a yearly basis, many developers feel the need to add additional functionality in order to entice users into upgrading the existing software. Unfortunately, the added features increase the size of the program and the system requirements needed to run it smoothly, eventually forcing the user to upgrade in order to run the latest software." (From: https://www.techopedia.com/definition/4237/bloatware, accessed 2 April 2016).

Almost humorously, the same article from Technopedia.com goes on to say: "Cloud-based, software as a service subscription models are seen as alternatives to bloatware because they reduce the need to resell products in the form of an annual update."

Au contraire. Subscription-based models of selling software have the effect of institutionalizing bloatware's first cousin, incrementalism, as software bugs, feature shortcomings, and expiring licenses all combine to drive customers to pay annually (or even more frequently) for a service that was once deemed a durable commodity. We now see incrementalism embedded in a "service" model, ensuring tech giants have their steady cash flow; tech companies first acclimated customers to annual updates (so customers could repeatedly secure the "latest and greatest") and then introduced those same customers to "software as a service."

125

This is not to say that good products cannot be outcomes of a software-as-a-service model, but we do note that savvy businessmen motivated by billions of dollars certainly are capable of including a tactic such as incrementalism in their arsenal of money-generating strategies. Likewise, governments clearly use incrementalism to further agendas related to centralized control. (See Martin Shapiro's "Judges as Liars" for a brief but well-written explanation of incrementalism at work in the realm of law and judicial proceedings. From http://scholarship.law.berkeley.edu/cgi/viewcontent.cgi?article=1265&context=facpubs, accessed 2 April 2016).

As the late Neil Postman noted many times, when confronted with a new technology (or new packaging of existing technologies), we should ask ourselves, "What is the human problem to which this technology addresses itself?" If we can't answer that question for ourselves with clarity, then we perhaps have no genuine need for the new technology.
Incrementalism, however, fogs our glasses when we look at Postman's question. If a vendor is selling a continual "service," then we cannot necessarily ask ourselves about any specific "problem" being addressed by a given technology. Software as a service grants us a vested right to define our problems via a future ongoing relationship with the vendor of that service. Indeed, we have altered both our points of view on our relationship to problem solving and the overall processes we employ to decide which tools we use to accomplish tasks. For an example of how radically we've changed, consider how (un)popular typewriters would have been had their vendors said, "they'll break a lot the first year, and some keys will be missing, but if you subscribe to use them as a service, we'll make sure you get an upgrade each year, so long as you keep the payments coming." This is the extent to which we have acquiesced by accepting rented word processor applications, and we got to this point by buying into one little software marketing change at a time—incrementally.

Lateral Surveillance

From http://www.surveillance-and-society.org/issue2(4)abstracts.htm:

1. Peer monitoring.
2. People watching other people.

Lateral surveillance is another term that we would hope to find in commonly used dictionaries, but the term is absent. We participate in lateral surveillance every time we heed the guidance of the US Department of

Homeland Security's "If you see something, say something" campaign. In a sense, labeling that set of behaviors for what it is—namely, state-induced peer-to-peer surveillance—unveils the call for tattling as more than mere bureaucratic bromides; widespread acceptance of lateral surveillance may be one of the most damaging examples of social engineering in an era full of them.

Human societies have seen lateral surveillance before, but today, digital technology, as we have created, implemented, and accepted it, is *the* single most powerful enabler for lateral surveillance.

A distinguishing feature of lateral surveillance is that it is meant to be implemented regardless of social or political standing. That is to say, common folk are not supposed to be influenced or deterred by social and political considerations when deciding whether to engage in lateral surveillance. So, for example, someone driving down a highway is supposed to feel free to make an anonymous phone call to report an erratic driver, regardless if the irregular driving originates with a thrill-seeking teenager or a drunk off-duty policeman. In fact, many likely believe the purest forms of lateral surveillance can be found in situations in which anonymity applies all around; citizens filming, taping, and reporting behaviors and identifying information such as license plate numbers or addresses (without any idea of the *person* behind that plate or address) lends a machinelike efficiency to state surveillance.

Some see lateral surveillance as a source of hope for a fairer and more socially balanced democracy. The thinking seems to be that a camcorder in hand while the police abuse their power is a sort of social "equalizer." We have seen instances like this (Rodney King comes to mind, of course), but we have also seen a sort of one-upmanship at work; police increasingly rely upon dashboard cams, and, more recently, vest-mounted cameras to "counter-document" events and incidents while on patrol.

In short, lateral surveillance may lead to useful derivative practices, such as sousveillance, and either of those practices may in isolated situations bring temporary help in equalizing the players in situations involving abuse of social, political, or even legal power. Nonetheless, neither lateral surveillance nor sousveillance can substitute for fair and balanced human relations. This is, in part, because lateral surveillance, by definition, entails distrust. Even more to the point, a society that embraces lateral surveillance in the spirit of Jeremy Bentham asks its citizens to institutionalize behaviors that explicitly demonstrate we do not deem one another worthy of living as truly free people.

Others may argue that ubiquitous state surveillance leaves average citizens little recourse other than lateral surveillance. The flaw in that argument is similar to the flaw seen when war protesters resort themselves to violence to get their point across. Engaging a flawed behavior using the same tools and means that form the essence of the flaw too often negates potential net benefit.

Mechanization of the Mind

Definition derived from
http://www.academia.edu/314119/Review_of_Dupuy_On_the_Origins_o
f_Cognitive_Science:

1. Reducing the activity of the human mind to machine-like processes.
2. Modeling the human mind solely on current understandings of physics and thereby using the abstraction (i.e., the mechanical model) to form or reform the original human mind itself.

Several authors in the last ten years have explored the idea that our attempts to humanize machines—the most obvious example being our ongoing efforts to turn computers into replicas of the human mind via artificial intelligence—may be backfiring. Steve Talbott and Sherry Turkle are among current thought leaders who question how we might be unwittingly impacting ourselves by adopting digital technology into every facet of our lives. They were among the first technology writers to follow up Joseph Weizenbaum's concerns about how artificial intelligence might impact the portions of our sanity that help us recognize the difference between the operations of a machine and life itself.

We believe Jean-Pierre Dupuy may have best articulated our concerns about the mechanization of the mind. Dupuy notes that as we focus on the abstraction of the human mind (in the form of a mechanical model), our "understanding" may be placing limitations on the actual human mind—i.e., we may be mistaking, inadvertently, the model's limitations as belonging to our own real minds. Dupuy quotes John Searle on this point:

John Searle, in his critique of cognitivism, has asserted that its principal error consists in confusing simulation and duplication. To quote one of his favorite examples, it would be absurd—as everyone will readily agree—to try to digest a pizza by running a computer program that simulates the biochemical processes that occur in the stomach of someone who actually digests a pizza. How is it then, Searle asks, that cognitivists do not see that it would be just as absurd to claim to be able to duplicate the neurobiological functioning of the mind by running a computer program that simulates, or models, this functioning? (From http://press.princeton.edu/chapters/i6920.html, accessed 26 March 2016).

Metaman

Definition derived from http://metanexus.net/essay/humans-plus-or-minus-introduction:

A [hypothetical] human being whose very nature has been improved through the use of applied science and other rational methods.

"Metaman," as we use it in MIAATQ, is an eponym. Variations of the term that we could use instead include at least the following: transhuman, MOSHs (Mostly Original Substrate Humans), post-human, cyborg, infrahuman, and cyberman. Several pundits see a significant distinction between some of these terms, as, for example, between "post-human" and "transhuman." For our purposes, any term that indicates a general belief in permanently "improved" technology-enhanced persons can be applied to our concern with *technology that has missed its mark via excess and/or misdirection.* In the case of any of the terms named above, the excess comes from a belief that human beings can—and should—*self-engineer.* [Editor's note: We recognize that even seeking to have an infected tooth extracted qualifies at some level as an inchoate form of self-engineering. We also recognize, however, that current efforts at mapping cognitive processes to specific areas of the brain, manipulating genetic codes at the most detailed levels, and seeking to emulate and exceed the very mind of man so that progeny will in effect be a new species—this is a qualitatively different level of self-engineering than we have ever known in the history of human beings.] We derive our specific definition for "metaman" from Langdon Winner's essay, "Are Humans Obsolete?," in which he quotes Gregory Stock's book *Metaman: The Merging of Humans and Machines into a Global Superorganism*:

Both society and the natural environment have previously undergone tumultuous changes, but the essence of being human has remained the same. Metaman, however, is on the verge of significantly altering human form and capacity….

As the nature of human beings begins to change, so too will concepts of what it means to be human. One day humans will be composite beings: part biological, part mechanical, part electronic....

By applying biological techniques to embryos and then to the reproductive process itself, Metaman will take control of human evolution...

Taking a few steps back (OK, *quite* a few steps back) from Stock's vision of Metaman, we note Nicholas Carr's very relevant comment about the vastly successful corporation, Google, and the assumptions and working milieu of its founders, Sergey Brin and Larry Page:

...their [Brin and Page's] easy assumption that we'd all 'be better off' if our brains were supplemented, or even replaced, by an artificial intelligence is unsettling. It suggests a belief that intelligence is the output of a mechanical process, a series of discrete steps that can be isolated, measured, and optimized. In Google's world, the world we enter when we go online, there's little place for the fuzziness of contemplation. Ambiguity is not an opening for insight but a bug to be fixed. The human brain is just an outdated computer that needs a faster processor and a bigger hard drive. [From Carr's seminal (yes, *seminal*) article in *The Atlantic*: "Is Google Making Us Stupid?"]

Natural Will

[See full definition of Gnomic Will, above.]

Panopticon

From http://www.yourdictionary.com/panopticon#americanheritage and the authors of MIAATQ:

1. Hypothetical prison proposed by Jeremy Bentham, having circular tiers of cells surrounding a central observation tower.
2. Manifestation of current mind set in society in favor of centralized control of human behavior via ubiquitous observation, in accordance with the theories of social control of Jeremy Bentham.

The design of a physical panopticon optimizes the ability to surveil. A panopticon, as Jeremy Bentham designed it, is a prison in which darkened guard towers, a centralized yard, and lighted prison cells enable authorities to monitor inmates (i.e., Bentham's design enabled top-down surveillance from prison leaders, both literally and figuratively). We argue that

Bentham's design and theories of social control relate to digital technology because today's increasingly pervasive mobile devices, social networks, and the emerging Internet of Things (IOT) each contribute to an infrastructure that amounts to a virtual panopticon. As more of our lives and services move to digital interfaces, we are implicitly required to allow continuous monitoring of our activities and communications.

Bentham was clear in his belief that the simple act of observation can become a powerful form of social control. As an ostensibly free people, we should consider the consequences if Bentham were right, because the amount of personal information we now share with the state (and with corporate quasi-agents of the state), wittingly or not, goes beyond the most optimistic estimates of effective surveillance Bentham's physical prison design could have generated.

In short, we believe few societies of any previous era would have allowed the vast amount of personal information to be collected that we routinely surrender today through our use of digital devices and technologies.

Participatory Panopticon

Definition derived from
http://www.worldchanging.com/archives/002651.html:

1. Constant surveillance done by the citizens themselves, by choice; a bottom-up version of the constantly watched society.
2. A combination of traditional surveillance, lateral surveillance, and sousveillance.

In a sense, the concept of the *participatory panopticon* is lateral surveillance in action, in full bloom. Jeremy Bentham introduced the prison-based concept of the panopticon itself, and much later, Foucault brought the concept new attention on the world stage. The panopticon, by the very definition provided by Bentham, its creator, calls for ubiquitous surveillance as a societal "good" for the purposes of control and social order. More concretely, we can say that in Bentham's thinking, we should extend to all adults the same 24/7 observation that caring parents provide for their newborns and toddlers; in short, Bentham would argue, by depriving us *absolutely* of any form of privacy, we should never grow collectively into any genuine sort of independence or self-reliance (with all the unpredictability and de-centralized action such a human condition implies).

With our recent acceptance of the *participatory* aspect of lateral surveillance, we see a twist to the already twisted conceptual foundations of the panopticon. We can assert that a social fabric that makes heroes of tech giants who say things such as, "We don't need you to type at all. We know where you are. We know where you've been. We can more or less know what you're thinking about,"—such a society has moved beyond acquiescing to top-down surveillance and has embraced lateral surveillance fully. If Eric Schmidt can say such things in public, and receive barely a whisper of backlash, let alone a boycott of his company's products, then we are saying by our inaction, yes, let's continue to acquire and use digital tools, applications, and devices that make what we're thinking the business of others. (See http://www.theregister.co.uk/2010/10/04/google_ericisms/, accessed 6 March 2016.)

Described using blunt and crude speech typical of modern America: a generation that willingly posts its underwear anywhere no longer embraces the dignity of privacy nor the value of mystery. Such a people have been mocked by workers at the NSA. (See http://www.zerohedge.com/news/2013-09-09/nsa-mocks-apples-zombie-customers-asks-your-target-using-blackberry-now-what).

Why mock? Sadly, the mentality seems to be that average citizens are—to quote someone at the NSA itself—"zombies," and zombies who willingly divulge every detail of their lives in the name of "sharing." And this is the saddest aspect of the participatory panopticon: that we are willing to embrace and bring to fruition *via our own actions* Bentham's stated goal that those living in the panopticon, whether they be prisoners or subordinates, should be under "irresistible control." Again, we participate in creating and maintaining Bentham's prison even by simply acquiescing to the digital tools and structures that form its electronic walls.
Quoting Tim Raynor: "There are no guards and no prisoners in Facebook's virtual Panopticon. We are both guards and prisoners, watching and implicitly judging one another as we share content." (From http://philosophyforchange.wordpress.com/2012/06/21/foucault-and-social-media-life-in-a-virtual-panopticon/, accessed 25 May 2013.)

Progress

From http://www.merriam-webster.com/:

Gradual betterment; the progressive development of humankind.

We include the simple word "progress" in our key definitions because it may be the most overused and ambiguous word in current writings about technology. Having analyzed how people use the word "progress" in many discussions involving technology, it seems that the common understanding of progress includes at least four assumptions, *all of which we should question*:

1.   Our society will integrate more technology with the passage of time; this trend never reverses.
2.   Increased use of technology equals a "more advanced" civilization.
3.   Increased use of technology is an overt expression and confirmation of the Whig theory of history (i.e., the passage of time equals the improvement of the world).
4.   Technology today is better for the lives and wellbeing of all people than technology was for all people living yesterday (i.e., technology just keeps getting better and better, with the definition of "better" remaining steadfastly undefined).

Ever-growing technology has not always been viewed as an unquestioned good. Ancient Greeks, for example, apparently had considerable ability to focus on technology, but chose not to do so. In other words, not all people of all eras have believed that technology is an unequivocal "good," nor have very many societies embraced belief that increasingly complex and/or ubiquitous technology is synonymous with "progress." Most of us in *this society* clearly do embrace such a belief.
As the late Neil Postman said, "In the culture we live in, technological innovation does not need to be justified, does not need to be explained. It is an end itself because most of us believe that technological innovation and human progress are exactly the same thing. Which of course is not so." (From a lecture Postman delivered at the College of DuPage on 11 March 1997, as transcribed from https://www.youtube.com/watch?v=hlrv7DIHllE, accessed 22 March 2016.)

Reification

Derived from www.dictionary.com/:

1. The conversion into or regard of a concept as a concrete thing.
2. To reify a concept.

We can arrive at a quick understanding of reification with a look at the following riddle: "It is the subject that presents itself with self-explanatory powers." Subjects, topics, elements of thought, and definitions (i.e., not humans, in and of themselves), do not, in the truest sense, ever explain anything; without a person to ruminate or discuss a subject, there is, by definition, nothing explained and nothing understood. Just as truly, we assert that subjects do not even present themselves; both the acts of presentation and explanation call for a *person* as the actor. And, that said, our riddle points to reification (or its first cousin, personification) as both the answer to the riddle and a chimerical construct. (OK—fair enough, "riddles" don't point. For the most part, only humans, great apes, and dogs point).

Karl Marx and Georg Lukacs famously used the concept of reification in their writings. Our interest in "reification," as we use it in MIAATQ, does not reference politics and sociology, per se. We discuss reification for the same reason we attempt to use the active voice as much as possible in our writing. We are concerned that in our era we've somehow found it simpler to talk, think, and act as if our own ideas and actions (and inventions) are not really our responsibility. We're not certain how much of this trend or phenomenon arises from our decisions to adopt and employ digital technology. Nonetheless, we join no less a light than Neil Postman in noting that quantification, the first and most essential ingredient of digital technology, begins with the temptation to reify. As Postman notes in his masterpiece, *Technopoly: The Surrender of Culture to Technology*: "The first problem is called reification, which means converting an abstract idea (mostly, a word) into a thing. In this context, reification works in the following way: We use the word 'intelligence' to refer to a variety of human capabilities of which we approve. There is no such thing as 'intelligence.' It is a word, not a thing, and a word of a very high order of abstraction. But if we believe it to be a thing like the pancreas or liver, then we will believe scientific procedures can locate it and measure it."[154]
Technology and intelligence measuring, however, are by no means the only fields in which people fall upon reification for refuge from responsibility. Martin Shapiro dares to unveil lawyers and judges as some of those most likely to reify, in this case, "the law":

Courts have decided, however, in all of the societies that have a modern judicial system, to avoid the appearance of deciding cases based on judicial whim. ... in all modern societies, and in all cases, judges tell the loser: 'You did not lose because we the judges chose that you should lose. You lost because the law required that you should lose.' That is the answer arrived at to satisfy the losers through hundreds of years of experiments in numerous societies. This paradox means that although every court makes law in a few of its cases, judges must always deny that they make law.[155]

As Shapiro and all lawyers and judges know, we're all familiar with the idea that "the law" and the Constitution of the US demand and require certain behaviors from us. This is an example of how reification can serve useful purposes in our society. The fact is that every time a judge uses the language Shapiro references (i.e., "You lost because the law required that you should lose"), the judge is really saying many, many things that cannot possibly fit into the time we allot for a typical court case. If the judge had to explain at each verdict how he'd gone to law school and studied various rulings that other lawyers had derived from other rulings made by other judges that had been written down and studied by various lawyers throughout many decades, if he had to explain the concepts of natural law and describe in detail where he found and interpreted various norms, our legal system would break down immediately for lack of time, if not for a deluge of boredom-induced vomitus. Shapiro knows this, and while he does not deny the usefulness of reification in our judicial system, he is honest and insightful enough to realize we can all benefit from recognizing those somewhat-hidden circumstances in which we hide behind reification.

This point brings us to the role of reification in digital technology. We'll call on another lawyer for insight into how reification in digital technology parallels reification in our legal system. Lawrence Lessig, author of *Code: And Other Laws of Cyberspace*, was among the first to recognize how *un-grudgingly* we've decided to first reify, and then by default, *personify* "cyberspace" so that the builders, programmers, and owners of computers and computer networks (aka, the Web, Internet, or my personal least-favorite, "the Cloud") are given both the blame and glory for the daily ebb and flow of our interactions with computers. We all have experienced the clerk or receptionist telling us he "can't do such and such" because "the computer won't let me." Lessig digs for and helps expose the roots of such pathetic groveling at the feet of technology:

There's a common belief that cyberspace cannot be regulated-that it is, in its very essence, immune from the government's (or anyone else's) control. *Code*, first published in 2000, argues that this belief is wrong. It is not in the nature of cyberspace to be unregulable; cyberspace has no 'nature.' It only has code—the software and hardware that make cyberspace what it is. That code can create a place of freedom—as the original architecture of the Net did—or a place of oppressive control. (From http://www.amazon.com/Code-Other-Laws-Cyberspace-Version/dp/0465039146, accessed 21 May 2016.)

Perhaps the most important point of Lessig's many profound arguments in *Code* is the one noted above, namely *that cyberspace has no nature*. Behind every bit of functionality in computers, computer networks, and "cyberspace" is someone, i.e., a *person* who wrote software code. No matter how abstracted our actions (and inactions) are in the presence of digital technology from

those who wrote the salient software code, human beings are responsible
for the resulting successes and failures we experience while engaging digital
technology.

In short, the more we make ourselves cognizant of the times and places we
reify and personify digital technology, the more likely we will be to make
intelligent and responsible choices about employing digital technology. Or
not.

Self-Censorship

From http://dictionary.cambridge.org/us/dictionary/english/self-
censorship:

*Control of what you say or do in order to avoid annoying or offending others, but without
being told officially that such control is necessary.*

Like many of the terms and phenomena we focus upon in MIAATQ, self-
censorship is not *solely* a new outcome or by-product triggered by the
widespread use of digital technology. Public opinion scholar Elisabeth
Noelle Neuman wrote about a key component of self-censorship, the
"Spiral of Silence," at least as early as 1974—long before "social media" or
the "NSA" were household terms. Neuman noted, "the climate of opinion
depends upon who talks and who keeps quiet" (See
http://thefederalist.com/2015/08/27/10-key-ways-to-break-the-mass-
delusion-machine/, accessed 13 October 2015). Neuman's point seems
intuitively grasped: if people know they cannot comfortably discuss a given
topic with friends, colleagues and associates, public opinion about that topic
is effectively censored.

As happens so often, however, digital technology serves to amplify certain
negative aspects of the already present phenomenon of self-censorship. For
example, Americans in the last few decades have read frequently about the
law enforcement technique of "profiling." Many stories have appeared in
local papers—long before anyone knew of Will Binney or Ed Snowden—
talking about FBI or other federal officers telling local police what type of
person they should look for as the perp in any given crime scenario. Details
provided often included race, gender, rough physical appearance, details
about the perp's personality and sociability, and even the type and color of
vehicle the perp would likely drive.

Even a casual reader could deduce that law enforcement was using banks of
data to associate certain collections of relatively small personal details in
order to understand and predict criminal behavior. "Profiling" became,

sadly, an all-too-familiar term in middle America. And this begs the question as to whether awareness of profiling motivates people to change their behaviors (and even their tastes and "style") in a reaction against profiling—regardless if these changes are made consciously or not. Likewise, early in the 21st century, Americans were so aware of having their online communications routinely surveilled that some even organized an event in which hundreds (or perhaps even thousands) intentionally texted and typed into emails terms of interest to federal agencies in an attempt to "jam" the government's ability to monitor cyberspace for keywords. The existence of such an event, whether or not it had its intended effect of "crashing" federal Web surveillance for the day, indicates a broad and deep awareness of "being watched" online. Fast forward to post-Snowden awareness, and we are all too familiar now with terms like "metadata" and "keyword search."

We could cite many studies showing the negative impacts of surveillance on human freedom and creativity. Karen Turner, writing for the *Washington Post*, summarized much of our current concern about unintended consequences of surveillance when she penned the article, "Mass Surveillance Silences Minority Opinions, According to Study" (From https://www.washingtonpost.com/news/the-switch/wp/2016/03/28/mass-surveillance-silences-minority-opinions-according-to-study/, accessed 29 March 2016). Turner's opening sentences:

A new study shows that knowledge of government surveillance causes people to self-censor their dissenting opinions online. The research offers a sobering look at the oft-touted 'democratizing' effect of social media and Internet access that [supposedly] bolsters minority opinion.
The study, published in Journalism and Mass Communication Quarterly, studied the effects of subtle reminders of mass surveillance on its subjects. The majority of participants reacted by suppressing opinions that they perceived to be in the minority.

Turner goes on to reference elements of the study that unmask the sad bromide asserting people shouldn't "care about online surveillance if they don't break any laws and don't have anything to hide." Turner notes comments from the study's lead researcher, Elizabeth Stoycheff: "She said that participants who shared the 'nothing to hide' belief, those who tended to support mass surveillance as necessary for national security, were the most likely to silence their minority opinions."
Stoycheff at one time focused on the perceived innate ability of the Web to promote democratic change. Sadly, her conclusions now include the following comment: "The adoption of surveillance techniques, by both the

government and private sectors, undermines the Internet's ability to serve as a neutral platform for honest and open deliberation. It begins to strip away the Internet's ability to serve as a venue for all voices, instead catering only to the most dominant." In other words, we now see self-censorship in action.

Situational Awareness

From the US Coast Guard manual at:
https://www.uscg.mil/auxiliary/training/tct/chap5.pdf:

The ability to identify, process, and comprehend the critical elements of information about what is happening to [fill in the blank].

A gut-level explanation of situational awareness may be to describe it as simply "being aware of the most important events and trends surrounding you as you undertake various tasks." In some ways, situational awareness might best be comprehended when we're around those who do not have it; young infants and elderly dementia patients come to mind. An infant cannot be expected to respond in any meaningful way to, for example, a fire alarm. Likewise, physicians and nurses often gauge the mental status of dementia patients by asking simple questions such as "what season of the year are we in?" or by asking elderly patients about the city or place in which they reside. In the case of both the infant and the dementia patient, awareness of surrounding circumstances and environments—the basic and general awareness we expect of all healthy adult human beings—is missing or in decline.

Needless to say, the surrounding circumstances and environment in the fields of medicine, aviation, and nuclear technology, to name a few fields, are far more critical to their practitioner's moment-to-moment wellbeing and survival than they are to most of us in our typical day-to-day lives. A wayward pin prick, a missed gauge reading or a valve at the wrong setting can ruin the day of any doctor, pilot, or engineer in those fields, and it can do so in a hurry. Technology, in those disciplines, escalates the potential for both good and bad events to unfold in almost any given situation.
In response to this heightened risk, many have tried to use yet another technology—information technology (which today means, for the most part, digital technology)—to augment situational awareness. Yet, results are mixed, at best.
The following three lines of text, culled from Jason Schappert's Mzero.com, exemplify what we see happening to situational awareness within field after

field. Taken verbatim from an advertisement for aviation aids, Schappert says:

"Technology Hasn't Changed the Way We Fly"

"Aerodynamic principles haven't changed. However, technology has added 'Increased situational awareness'"

"This increased situational awareness however can lead to cockpit distractions and too much 'head down time.'" (See https://digital-flying.s3.amazonaws.com/Digital-Flying.pdf, accessed 9 April 2016.)

In the innocuous statements above, Schappert captures, inadvertently, some of the irony and unintended consequences of using digital technology to try to *increase* situational awareness. The irony often plays out in this simple pattern: We recognize a problem newly introduced by technology (i.e., other than Daedalus and Icarus, no one worried about flying into the side of a fog-shrouded mountain until about 100 years ago), and then we try to apply various forms of digital technology to eradicate the first problem; i.e., technology solves a problem and, in the process, creates a new problem that calls for more technology as a "fix."

Of all the counter-intuitive impacts of digital technology, *decreased* situational awareness may be the most pronounced in terms of contrast in planned outcomes versus the actual real-world experience. As Schappert notes in his advertisement, people have developed marvelous tools to help pilots navigate through the most obscene weather and flight conditions imaginable. If, as a pilot, I can use radar to warn of a mountain or thunderstorm in my path, I've used digital technology to augment my situational awareness.

At the same time, the ever-crucial possibility for ambiguation enters the scenario. As Schappert notes, pilots who are "heads down," by definition, are pilots with decreased situational awareness about what is in front of them. Whether the technology the pilot is using is a printed map or the latest iPad loaded with aviation apps and strapped to the pilot's leg, we have no technology that can (or should) substitute for human attention to the flight path and the relevant aviation conditions.

We often see a knee-jerk reaction to the introduction of new digital technology; too many of us seem to anticipate expanded situational awareness as a given with each new technological "advance." Yes, the ability of modern radar to see through fog and project the resulting image onto a screen gives pilots "eyes" they could never have without that technology. So the danger to a pilot's situational awareness comes not from the new technology itself but from the way in which it is used; even prior to utilization decisions, as we develop and deploy any given new technology, a type of danger seeds in our misunderstandings about what technology really brings to humanity. Using machines to impose our will as we fly through foggy valleys solves a specific set of problems on any given flight, but it does not provide permanent improvement to *ourselves*.

Social Engineering

From http://dictionary.sensagent.com/SOCIAL ENGINEERING POLITICAL SCIENCE/en-en/:

Social engineering is a discipline in political science that refers to efforts to influence popular attitudes and social behaviors on a large scale, whether by governments or private groups. In the political arena, the counterpart of social engineering is political engineering.

[Editor's note: "Social engineering," as we use it in MIAATQ, has nothing to do at all with the "hijacked" term of the same name used by some to refer to a practice of hacking into software systems via confidence games.]

We follow Karl Popper's lead here and recognize two forms of social engineering: piecemeal and utopian. And, like Popper, we see the crucial difference between the two approaches as resting in the underlying attitude of those invoking social engineering.

Leaders must often consider the long-term consequences and outcomes of laws, mores, and social movements. We acknowledge this fairly obvious point and recognize that many decisions of state go far beyond the relatively straightforward implications of, say, deciding whether an intersection near a kindergarten needs a traffic light or a stop sign. Given that, we also understand that some leaders will at times think of themselves as "engineers" of society, responsible for choices that are much more complex and consequential than decisions such as where to build a schoolhouse or even where to deploy a weapons system, or other similar decisions that appear, at least initially, to be about strictly material matters Sometimes the actions of more traditional engineers—say, highway engineers or architects—have ended up influencing society in unanticipated ways. For example, when leaders in the Eisenhower administration, leaning on some of the military lessons of WWII, decided that our country could benefit from a nationwide network of multi-lane highways, those leaders hopefully understood that such a network would have massive costs in addition to its obvious benefits. We can't say with certainty that the US senators and congressmen of the 1950s realized that blessing a nation with coast-to-coast interstates would result in the demise of large portions of the country's existing public transportation system (e.g., local passenger trains and trolley systems); nor do we know if those leaders could foresee that social and consumer homogenization, resulting from easy highway access, would lead to vast networks of "chain" restaurants, so that people would end up routinely buying exactly the same hamburger in Carmel, California, that can be bought at a fast-food shop in, say, Martha's Vineyard on the East Coast. That is to say, we acknowledge that leadership decisions and events, even within (or perhaps *especially* within) a democracy or a republic, will often impact the long-term future of our society and culture.

When leaders *do* recognize that certain laws and political actions will bring potential long-term impact, we normally expect them to expose these impacts for consideration and discussion among the public. Neil Postman, for example, in some of his lectures and writings, brought up the example of the internal combustion-powered automobile. He asked the hypothetical question about rolling the clock back to 1911 and putting a plebiscite before the American public, in which we honestly discussed the air pollution and traffic congestion automobiles would bring along with their obvious and massive benefits. Postman indicates that had we floated such a plebiscite, he believes the American public would have still said, quote, "Shit, let's do it anyway" [i.e., make automobiles]. He then, however, makes the point that we might have also had some discussions about what we could and should do to minimize negative impacts.

Postman's hypothetical plebiscite and public discussion about cars provides an example of *piecemeal* social engineering. The automobile, we now know, brought with it a massive impact to our society, our culture, and even our landscapes. Even today, 100-plus years removed from the invention of the car, we see leaders of large cities limiting the times and places cars may travel and even limiting the size of cars used and the numbers of passengers carried in certain lanes of traffic. All of this in an effort to *openly* "engineer" human behaviors in a way that is meant to lead to improved living conditions.

By contrast, those leaders drawn to utopian social engineering too often seem enamored with the public's silent *manipulation*. Somehow, those looking for the proverbial "immanentization of the eschaton"[156] seem to believe the feats of social engineering required to bring it about should remain hidden from public review and criticism. It's as if the great unwashed dare not be active participants in planning their own cleansing. As to the realm of digital technology, we offer an example of an alleged *lack* of utopian social engineering in the form of the following quotation from

Bruce Schneier's ominously titled article, "We're Sleepwalking Towards Digital Disaster and Are too Dumb to Stop." Schneier notes, "The world-sized web will change everything. It will cause more real-world consequences, has fewer off switches, and gives more power to the powerful. It's less being designed than created and it's coming with no forethought or planning. And most people are unaware that it's coming." Schneier says, clearly, that he sees a lack of planning in the digital world—and when we think of utopia, lack of planning is rarely, if ever, a feature that comes to mind. What if, however, some leaders believe that social control *is synonymous* with utopian thinking? Marshall McLuhan famously said, "the medium is the message." We have in our midst, perhaps, those who believe that "the infrastructure is the message." If you believe (as many do) that the computer is the device of command and control,[157] and if some leaders see control as being an inherent good (as many leaders do), then it's not much of a stretch to believe we'll encounter those interested in promoting widespread use of digital technology in as many venues as possible—simply because those promoting it believe digital technology brings inherent ability to control behavior. The infrastructure, in this case, is no longer a means to an end (i.e., the "end" being peaceful citizens free to pursue their interests with discretion and success), but the infrastructure becomes the end in and of itself.

We note that somewhat forcefully moving one human task after another to the digital format will lead in many cases to major disruption and, in some situations, even real-life disasters. For those interested in control as an end, however, the utopian vision of a society that can be run (eventually) as predictably as a machine appears to outweigh any and all temporary costs. How powerful is the lure of the "utopian" vision of unfailing social control? Here in "the land of the free and the brave," some of our leaders have made choices about control that led the NSA's 32-year veteran, William Binney, to note that "we are now in a police state." He describes one of the processes leading to his conclusion as follows:

In fact, they [NSA-generated guidelines] instruct that none of the NSA data is referred to in courts – because it has been acquired without a warrant. So, they have to do a 'Parallel Construction' and not tell the courts or prosecution or defense the [the source of the] original data used to arrest people. This I call: a 'planned programmed perjury policy' directed by US law enforcement.[158]

Binney's colleagues swore to uphold and defend the Constitution, just as he did. With so many of his colleagues enabling the police state, Binney describes how powerful, then, must be the lure and appeal to believe that "social engineering" built upon fabrication and quiet manipulation is a greater good than even the esteemed foundations of our written (and openly agreed upon) Constitution. Indeed, digital technology appears, at times, to bring inherent appeal to people who think and operate in the control-centric manner traditionally attributed to utopian social engineers.

Social-Techno Metacognition

From the authors of MIAATQ, combined with the definition of Metacognition from http://www.merriam-webster.com/:

Awareness or analysis of one's own learning or thinking processes and attitudes regarding the relationship and interplay between society and technology.

 Given that we coined this term, we'll begin explaining it with a single example of the runaway impact of digital technology in our daily reading habits; as we write this book, the US, and indeed much of the world, is in the midst of exchanging centuries-old printed newspapers for Internet-based news and alternate news sites. This outwardly simple change comes with extreme and not easily noticed consequences.

Those who have engaged in careers requiring constant vigilance in the face of multi-faceted dangers understand the need for situational awareness; if your every mistake can lead to a fiery crash or loss of life, your career comes with built-in motivation to think at all times about both what you are doing and what those around you are doing as well. Military aviation and certain types of police work come to mind as being in that category of career.

Those engaged in more introverted or "quiet" activities, such as artists, many types of craftsmen, and categories of laborers, clerks, and bureaucrats, likely have less built-in motivation for constant vigilance. A clerk filing papers, for example, does not typically fear for his life as he opens a filing cabinet.

What happens, however, when the online newspaper the clerk reads every day is hosted on a site filled with technology that "reads us" while we peruse, logging our topics of interest, reading habits, and peccadillos, and then sharing that information around the globe? Or, for that matter, what happens when the emails we send and receive or the files we download are designated as items of "national security?"

These questions suggest that our accumulating mistakes of non-action and misjudgment about the technologies we deploy within our society are leading us possibly to a form of electronic prison the likes of which the world has never seen. Hyperbole? Alarmism? Fifty years ago, our mothers and fathers could not have conceived of continuing to call our nation "free" if their very reading habits were commodities to be exchanged in the interest of either commerce or faux security. When we abandon concepts of privacy, when we grant all things digital a status they do not deserve, when we simply fail to think deeply about the impact of the devices, tools, applications, and information infrastructures we put in place, then we, without the slightest exaggeration, open the door for extreme consequences.

To be more specific, and to expand our discussion about the current threat of digital technology well beyond the scope of our daily reading habits: When horrific technologies such as autonomous killing robots and self-targeting drones are under development, and when those who helped build the very technologies and infrastructures that enable these monstrosities to flourish are sounding the alarm (see recent warnings of Bill Gates, Eric Schmidt, Elon Musk, etc.), we are already reaping the consequences of previous ill-advised decisions—and, more importantly, of our *non-decisions* or *inaction*.

Hence, our call for social-techno metacognition. Most of us have allowed ourselves to be sold the proverbial "bill of goods" that digital technology *always* improves life. The most straightforward way to begin examining that dubious assertion, whether or not we do so via heuristics or via more formal indexing and testing of the impacts and consequences of employing digital technology, is to make a very conscious attempt to "open our eyes" to the real impacts of our choices regarding new digital tools, applications, and technologies. The veracity of our findings can only improve if we make a conscious effort to examine as many facets of the relationship between society and technology as possible. In short, both everyday users of "smart phones" and dedicated sociologists studying newly found maladies such as digital dementia should benefit from improving their social-techno metacognition.

Solutionism

From: Morozov, E. (2013). The Perils of Perfection. Retrieved February 22, 2016, from http://www.nytimes.com/2013/03/03/opinion/sunday/the-perils-of-perfection.html?_r=0

An intellectual pathology that recognizes problems as problems based on just one criterion: whether they are "solvable" with a nice and clean technological solution at our disposal. Thus, forgetting and inconsistency become "problems" simply because we have the tools to get rid of them—and not because we've weighed all the philosophical pros and cons [of doing so].

In MIAATQ, we argue that adopting a mindset that "if it can be done, then it should be done" is not the best way to increase digital reliability. We often find that mindset when we adopt an approach to life that Evgeny Morozov labels as "solutionism," a term that describes an engine governments and corporations use to drive adoption of many of today's digital technologies. In fact, solutionism may bury the root cause of societal problems, introducing new issues that can only to be "solved" again via—you guessed it—more technology. And it is this penchant for self-perpetuating "need/solution creation" that is the most alarming aspect of solutionism; i.e., if we routinely introduce new digitally based "solutions" without having first explored whether the *subtraction* of technology will solve any given problem, then we open the door to a self-feeding cycle of technology adoption. And we can ask, quite fairly, if engaging in such a cycle induces us to lose sight of the original problem to be solved?

Morozov summarizes the root of this cycle as succinctly as possible in the postscript to his *To Save Everything, Click Here*: When we choose to view digital technology as the *explanans* rather than accepting its true status as the *explanadum*, we grant license to those with a penchant for puzzle solving an opportunity to insert their puzzles in our lives as faux problems/solutions. Morozov quotes Gilles Paquet: "'Solutionism' [interprets] issues as puzzles to which there is a solution, rather than problems to which there may be a response."

"Puzzle solving" may sound harmless enough, yet the term signifies subtle but vitally important differences in our understanding of how we should approach life. We encounter countless decisions with minor ethical, moral, and social implications on a daily basis. Do I stop by to visit my aging parent, or do I drop by the local pub for a brew? Do I send my daughter to public school or choose homeschooling? Do I monitor my children's selection of TV shows or chalk it up as simply "kid's stuff?" Each of these decisions, small enough in and of themselves, constitute life when viewed collectively. Amazingly, many, many digital technologists today would like to quantify the various elements of these decisions and then provide us with "apps" to ease the burden of choosing the right set of actions from our human hearts. In doing so, they relegate consequential aspects of daily living to the status of puzzles.

So, we should ask what happens when we delegate our lives to apps, i.e. to digital technology? The real consequences may take years to reveal themselves; I know of an otherwise sharp financial analyst who admits she cannot do relatively simple arithmetic in her head. Given how ubiquitous digital calculators are, I didn't notice her missing arithmetic skills for the longest time, and maybe little harm done, in that particular case. Moving up the ladder of consequences, it's becoming increasingly evident that many people are losing their ability to read maps. GPS, with its wonderfully simple step-by-step instructions, renders map reading a nearly useless skillset for an increasing number of drivers. What happens when the GPS quits? If your answer to that question is "We'll engineer ways to ensure GPS *never does quit*," then you are already likely a solutionist; map making and map reading are ancient skills that served the entire globe quite well for millennia. If, while looking to solve one of life's problems, existing simpler technologies such as maps don't even enter our problem-solving processes, then we've almost certainly embraced a self-feeding cycle of technology acquisition—which is to say, we've removed our primary focus from the real-life problem and moved that focus to the faux solution.

The world may not end because we can no longer read maps (and thereby give up the ability to understand the larger picture of where we are on the globe at any given moment), but adopting technologies that lead to the loss of our children's ability to think—that malady, known as *digital dementia*, may just end our world as we know it. We allowed our children to use digital calculators, desktop computers, digital tablets, "smart" phones, and a long parade of other digital tools for many years before we began to notice that the societies with the most access to digital tools and toys were filling up with young people who could no longer complete tasks as simple as remembering their own telephone numbers. Medical researchers have now joined writers such as Nicholas Carr in sounding the alarm: we may, indeed, destroy our own ability to think deeply and clearly by embracing too many digital tools in too many situations.

Solutionism, the act of focusing on the solution to the loss of bearing in regard to the original problem, does have consequences.

Sousveillance

From yourdictionary.com:

The recording of an activity from the perspective of a participant in the activity [as

146

opposed to surveillance, which describes recording of an activity from the perspective of an observer, monitor, controller, or regulator].

"Sousveillance" is a neologism meaning surveillance from the bottom up (i.e., from the bottom up of a society, institution, or enterprise). Sousveillance and lateral surveillance both share a reliance on a reversed social order in that "normal" surveillance typically begins at a higher level of any given social hierarchy and looks down; e.g., guards surveilling prisoners, bank officials surveilling tellers, etc. Sousveillance, by definition, involves those at the lower levels of a given social structure "looking up" to observe, and at times, expose behaviors of their "betters," typically for purposes of activism and/or socio-political change.

In the US, the Rodney King incident provides an early and vivid example of technology-based sousveillance. More recently, Julian Assange and Edward Snowden have taken sousveillance to new extremes. (Time will tell if their actions are simply a part of a Hegelian dialectic that will result in even more broadly "accepted" surveillance.)

"Sousveillance," the word, is new, but real-life approximations of the concept itself are not. Jeremy Bentham wanted surveillance to flow in all directions of social structures. That said, the idea (and capability) of having those in the lower classes looking explicitly "upward" in a social structure is likely an important departure from the mindset of Bentham and most generations before the 20th century (and the invention of miniaturized cameras and recording devices). Bentham and his ilk believed in the power of controlled and concealed observation used to maintain social order. So, for example, while he accepted (and encouraged) surveillance of prison guards, it seems that the direction of surveillance was understood to always point downwards within any given institution or society.

Sousveillance, often praised as a sort of technology-based savior for egalitarianism and democracy, does have its risks and liabilities. Governments and institutional leaders are very likely to react to sousveillance by *increasing and/or augmenting* their own surveillance. Recent examples include policeman now carrying cameras in both their vehicles and on their persons to record exchanges with citizens. Likewise, accepting sousveillance as the common person's response to excessive government surveillance ironically increases practices of silently watching others in lieu of engaging them directly; i.e., when we accept sousveillance as a legitimate social response, do we make ourselves part of a movement away from mutual trust and decency into the arms of intolerance and power-based social control?

Teleology

From http://www.merriam-webster.com/:

The fact or character attributed to nature or natural processes of being directed toward an end or shaped by a purpose.

For our purposes in MIAATQ, "teleology" is roughly the equivalent of the popular phrase, "the meaning of life." Some disciplines include discussion of teleology very often and very directly (philosophy, theology), while others seem to encompass a less visible teleology within their unstated philosophies; examples might include "wellbeing" as the teleology of psychology and "stability" (or, sadly and increasingly in our time, "control") as the teleology of politics.

Teleology becomes important in a book about digital reliability when one considers a growing group of highly educated and academically influential writers openly talking about transforming the very core of what it means to be a human being. This change they propose (or deem as inevitable) via the combination of artificial intelligence and genetic engineering. And, of course, genetic engineering in post WWII is far beyond the pea-picking methods introduced by Mendel; serious genetic engineering today relies upon digital technology in just about every way.

Perhaps the most salient reasons for considering teleology in MIAATQ are evident when we include in our discussion determinism, historicism, and the Whig theory of history. These terms implicitly contain a pronounced secularism that denies all forms of traditional belief in a personally involved deity. In this "belief vacuum," someone or something sentient must intervene if we are to admit the existence of a teleology of any kind; i.e., by definition, "meaning" does not appear out of thin air.

Our concern, then, is that unless the human beings frantically building digital technology recognize and admit their own teleologies—whether those teleologies be explicit, implied, hidden, religious, secular, dastardly, or even completely benign in nature—then we have the horrifying potential for unrecognized idolatry of human will. The wars and totalitarianism of the 20th century certainly should give us pause when we consider the shadows cast by unconstrained adulation of what man can do using science to impose human will on nature (and fellow human beings).

Transhumanism

From
http://www.oxforddictionaries.com/us/definition/american_english/trans
humanism:

The belief or theory that the human race can evolve beyond its current physical and mental limitations, especially by means of science and technology [See also the full definition of Metaman].

Whig Theory of History

From https://mises.org/library/progressive-theory-history:

The Whig theory of history says that history is an inevitable march upward into the light. In other words, step by step, the world always progresses, and this progress is inevitable.

The various labels "Whig history," "Whig theory of history," "Whig interpretation of history," etc., share an etymology rooted in Herbert Butterfield's 1931 book, *The Whig Interpretation of History* (see reference at the end of this paragraph). Butterfield, in his day, was familiar with Whigs as a British political party roughly interested in furthering all things parliamentarian (as opposed to Tories, who generally favored all things monarchical). Butterfield believed that the Whig writings of his era indicated a too ebullient optimism that constitutional monarchy comprised the be all and end all of political and social development. This excess of optimism about the "rightness" of their own political system led to, in Butterfield's words, an approach to history that "...studies the past with reference to the present ... Through this system of immediate reference to the present day, historical personages can easily and irresistibly be classed into the men who furthered progress and the men who tried to hinder it; so that a handy rule of thumb exists by which the historian can select and reject, and can make his points of emphasis" (See http://www.eliohs.unifi.it/testi/900/butterfield/chap_2.html, accessed 13 February 2016).

Modern optimists, many of whom share belief systems rooted in the Whig theory of history, tend to trumpet an ever-better future (i.e., endless progress) based on hope in science and technology. This may be because many modern scientists believe the scientific method provides an error-free

149

lens with which to view the past.

Yet, there's a problem with attributing perfect objectivity and, in some cases, infallibility to the processes and methods of one's belief system (e.g., the scientific method), and the root of that problem can be summarized in One quotation from Søren Kierkegaard: "Life can only be understood backwards; but it must be lived forwards."

As we observe future events unfold in real-time—which, not being omniscient nor even prescient, is our only choice—we react using our *gnomic will*; i.e., we take action based on what we think we see happening around us. Having done so, when we have time to reflect on our past actions (and those of others), we have the opportunity to impose interpretation on those actions. The huge problem we create for ourselves comes when we grant a completely unmerited "rightness" to our necessarily subjective interpretations of past belief systems and actions.

Indeed, Karl Popper argued that all observation is from a point of view. This is a pivotal assertion, as, in turn, our interpretations of the results of our past employment of the scientific method rely upon our observations. This point is a step removed from simply saying that "scientific method relies upon observation." In this case, we're saying that judging the rightness of past "accomplishments" of science depends also upon observation, specifically upon observing those accomplishments by, as Kierkegaard would have stated it, "looking backwards" while continuing to "live forward." Thus, unless we're capable of complete selflessness while "living forward," we will always be burdened with the temptation to bring our own lenses and filters to our observations and interpretations of the past.

To provide a concrete example in the realm of digital technology, cyber-skeptic literature is currently filled with references to "digital dementia" and other negative impacts to the health of young people who rely too much upon digital devices (e.g., cell phones, gaming devices, laptops) in their everyday living. Concurrent with these concerns, Alphabet (Google) and other tech companies are pounding out driverless cars and trucks, simultaneously wrangling with politicians and public opinion to make these cars/robots a near-future reality for all of us. If the leaders and engineers of these companies adhere, consciously or otherwise, to beliefs that "technological advances" are both inevitable and good, then they are more likely to miss the valuable cautionary lesson implicit in the recent identification of digital dementia as a destructive side effect of digital technology. More specifically, it's fair to focus our concern about self-driving cars even more sharply; we can ask whether or not concentrating on the very real concern about the physical safety of self-driving cars actually obscures the potentially far-broader impacts that widespread use of self-driving cars would have on the psychology and situational awareness of millions of drivers?

In short, the term "Whig theory of history," as we use it in MIAATQ, shares with *historicism* an inherent excuse to claim—falsely—that we can possess an unerring objectivity about past events and future trends. When we operate under a pretense of objectivity, an objectivity that even hints that we possess a type of selflessness that, frankly, does not exist in human beings as we know them, then our ability to be deceived about the real consequences of our attitudes, choices, and actions grows seemingly without bounds. We should be able to notice, for example, that embracing yesterday's digital devices and wizardry has provided us certain forms of convenience at the terrible cost of damaging our children's minds.
As we study the reliability of digital technology, particularly as we study it within a milieu of the ever-welcoming arms of our media, government, corporations, and, at times, all of society, we need to remain aware of our own implicit filters; if, as Butterfield suggested, we have succumbed to the temptation to continually judge the past based on what we have today, then we will likely miss opportunities to learn genuinely from our past mistakes.

FOOTNOTES

[1] Jon Ronson. "How One Stupid Tweet Blew Up Justine Sacco's Life." *The New York Times*. 12 February 2015. http://www.nytimes.com/2015/02/15/magazine/how-one-stupid-tweet-ruined-justine-saccos-life.html

[2] Ben Fritz. "Hack of Amy Pascal Emails at Sony Pictures Stuns Industry." *The Wall Street Journal*. 11 December 2014. http://wsj.com/articles/hack-of-amy-pascal-emails-at-sony-pictures-stuns-industry-1418259456

[3] The Greek myth of Daedalus and Icarus certainly predates Mary Shelley's work by a millennia or three, and melted wings leading to drowning qualifies as a tech failure. Shelley, however, wrote *Frankenstein* in 1818, at the beginning of the West's passionate and frenzied fling with all things technological. In addition to the wonderful timing of her novel, Shelley is incredibly prescient in writing about a technology-created being more than 100 years in advance of successful organ transplants, artificial intelligence, and genetic manipulation.

[4] Langdon Winner. *Autonomous Technology* (1977), p. 315.

[5] Langdon Winner. *Autonomous Technology* (1977), p. 312.

[6] Jeffrey Herf. "Technology, Reification, and Romanticism." *New German Critique*, no. 12, Autumn 1977.

[7] Ed Snowden, in video accessed 15 April 2016 at: http://www.theguardian.com/music/2016/apr/15/jean-michel-jarre-records-with-edward-snowden-nsa-whistleblower

[8] See our "Self-Censorship" entry in our Table of Concepts.

[9] Peggy Noonan. "What We Lose If We Give Up Privacy." *The Wall Street Journal*. 16 August 2013. http://online.wsj.com/article/SB10001424127887323639704579015101857760922.html?mod=WSJ_hpp_sections_opinion

[10] Jeffrey Herf. "Technology, Reification and Romanticism." *New German Critique*, no. 12, Autumn 1977, p. 188.

[11] Jeffrey Herf. "Technology, Reification and Romanticism." *New German Critique*, no. 12, Autumn 1977, p. 188.

[12] We discuss Morozov's points of view in more detail later; here we use his phrase "Internet solutionism" as a pole in a continuum of attitudes toward digital technology because it reflects the extremism he mocks in the title of his book, *Click Here to Save Everything*. In that book, he makes clear his belief that too many of us today act as if we truly believe every single problem of mankind can be addressed via the digital.

[13] One of my writing mentors, the late Jack Williamson, who I always

deemed to be both a master of science fiction writing and a level-headed thinker—traits that don't always coexist—wrote, among the dozens of books he penned, a novel that explores what appears to a be a dystopian view of a future filled with autonomous technology in the form of forcefully helpful robots. The human protagonist of his story evades these robots throughout the novel, apparently viewing them as the very embodiment of technology turned explicitly evil. To my surprise, Williamson ends the story by showing that submission to life run by these robots can actually be pleasant and even fulfilling; apparently the late Williamson believed, at least when he wrote that novel, that only the protagonist's quasi-Luddite attitude was at fault. With this plot twist, Williamson—at least in *one* of his many, many novels—joins other "transhumanists," such as Google's Ray Kurzweil, who defends the "advancement" of humanity via machine-based mimicry and enhancement. See the article at http://en.wikipedia.org/wiki/Transhumanism for an introduction to the concept.

[14] Some in the US media even seem willing to prepare the way for public acceptance of autonomous technology. "Why America Wants Drones That Can Kill Without Humans" by Joshua Foust (yes, a nearly tremendous irony aside, except for the spelling, the author's last name is "Foust"), discusses "lethal autonomous robots," or LARS, that sound like elements of a Schwarzenegger movie. Giving this article, published in DefenseOne.com, such a title is for us a point of contention. We are part of America and have absolutely no desire for LARS to be developed, thought about, deployed, or used. A common technique of many journalists and editors in our media seems to be to prepare the soil for radical changes in technology by wording articles from an inherently deterministic viewpoint, i.e., why bother to sound an alarm—and trigger much-needed debate—over a clearly radical change in how we use (and abuse) technology when you can simply use a bit of social engineering to prep public opinion for its acceptance? Foust's article can be accessed at http://www.defenseone.com/technology/2013/10/ready-lethal-autonomous-robot-drones/71492/

[15] See John Taylor Gatto's "The Underground History of American Education" for an excellent discussion of the negative impact of our public school systems on the ability of citizens to see, hear, and understand the human needs in our colleagues and neighbors.

[16] Donald MacKenzie and Judy Wajcman (eds.). *The Social Shaping of Technology* (Open University Press, 1985), p. 8.

[17] Daniel Chandler. "Technological or Media Determinism." 18 September 1985. http://www.aber.ac.uk/media/Documents/tecdet/tdet05.html

[18] See article by Justin Maiman on *Daily Ticker* by *Yahoo Finance*. Video: Aaron Task interviewing Intel's Brian David Johnson.

http://finance.yahoo.com/blogs/daily-ticker/smart-toasters-really-130854008.html

[19] At the time of publication of MIAATQ, we do not know if WalMart is using RFID technology in the retail items that go out its doors. For a brief history of WalMart's work with RFID in its supply chain and even on its retail shelves, see "RFID News: Will WalMart Get RFID Right This Time?" in *Supply Chain Digest*, http://scdigest.com/assets/On_Target/10-07-28-1.php?cid=3609

[20] See: http://www.zebra.com/us/en/solutions/getting-started/rfid-printing-encoding/rfid-basics.html for basics about RFID. As to transmission of a retail customer's location—this is not typical functionality in an RFID chip. The chip, however, does not need to actively broadcast location information for the location of a garment (or any item) carrying an RFID chip to be determined. Current RFID technology requires close proximity of the RFID detecting device. Thus, an RFID chip identified at a store on 100 Main St. can be reasonably assigned that location, along with time and date information, in a database tracking it. The information trail this leaves—purchases made with credit cards are easily linked to a buyer's name—provides corporations or governments the potential to "follow" the movements of consumers by simply combining streams of data from credit card companies and RFID scanners. Whether or not this is the intent of those deploying RFID chips is more or less the crux of our point here, the use of digital technology without adequate forethought and debate often leads to unintended (and sometimes *unforeseeable*) outcomes, even possibilities of the Pandora's box variety.

[21] See http://www.chron.com/news/medical/article/Robot-hot-among-surgeons-but-FDA-taking-fresh-look-4419667.php

[22] See http://www.chron.com/news/medical/article/Robot-hot-among-surgeons-but-FDA-taking-fresh-look-4419667.php

[23] Stephen L. Talbott. "The Fundamental Deceit of Technology." http://netfuture.org/1995/Dec1495_1.html#3

[24] We discuss Hegelianism more in detail later; the reference here is to the fact that so often today when someone points out a failure in digital technology—even a failure of catastrophic proportions—only in the rarest of cases does someone suggest use of a remediating process *that does not* include digital technology. The well-worn pattern seems to be: 1) identify a use of digital technology that failed; 2) describe in detail how bad the consequences of that failure are; 3) call for expanded use of "improved" digital technology to solve the original problem.

[25] Jiantao Pan. *Software Reliability* (Carnegie Mellon University, Spring 1999). http://www.ece.cmu.edu/~koopman/des_s99/sw_reliability/

[26] Gautam Naik. "Storing Digital Data in DNA." *The Wall Street Journal*. 24 January 2013.

http://online.wsj.com/article/SB100014241278873245393045782598835075
543150.html

[27] Jiantao Pan. *Software Reliability* (Carnegie Mellon University, Spring 1999). http://www.ece.cmu.edu/~koopman/des_s99/sw_reliability/

[28] Using digital technology impacts situational awareness (SA), a concept used throughout aviation. While my SA improved significantly when I first flew an aircraft with a digital readout of a radar-provided, above-ground-level altitude, other aspects of my SA degraded as map checking and primary instrument scans became less important (so long as everything digital was in working order).

[29] Pilots and others who work in the night need to maintain "visual purple" in their eyes in order to keep their vision as ready for low-light conditions as possible. Exposure to light—sunlight or even bright screens in a cockpit—temporarily depletes visual purple and leaves the pilot's eyes with greatly reduced ability to see dimly-lit ground features, darkened cockpit features, or even maps. See the article at http://en.wikipedia.org/wiki/Visual_purple for more information. We also note that as this book is being written, talk of "driverless" cars is in all the press. In the various articles about these incipient tech wonders, so far no one has mentioned the factors discussed here: the withering of vital human driving skills that will inevitably follow when passengers repeatedly allow "the computer" to drive.

[30] We dedicate Chapter 4 to defining and discussing in detail tech-aided ambiguation of roles. The issue is one of several "uber issues" affecting our lives and can be seen in many of the real-life anecdotes and tragedies we cover in other chapters.

[31] Evgeny Morozov. *Click Here to Save Everything* (PublicAffairs Books, 2013).

[32] Morozov, p. 49.

[33] Joseph Weizenbaum's work with computers as far back as the 1960s led to multiple professionals in the field of psychology to make early knee-jerk reactions in the direction of using computers to replace human psychologists.

[34] See Morozov, pp. 79-85, for an excellent discussion about how Internet-based transparency impacts behaviors of the observed.

[35] In "New MoD Report Extends Vision to 2040" (http://www.oldthinkernews.com/2010/03/new-mod-strategic-report-extends-vision-to-2040/), Daniel Taylor quotes the UK MoD paper as saying "it may become difficult to 'turn the outside world off,' and …'Even amongst those who make an explicit life-style choice to remain detached, choosing to be disconnected may be considered suspicious behaviour.'"

[36] See Stephen L. Talbott's marvelous "Remembering Ourselves," which is the introduction to his book, *Devices of the Soul: Battling for Our Selves in an Age of Machines*, O'Reilly Media, April 2007. "Remembering Ourselves" may be

found at http://natureinstitute.org/pub/ic/ic17/devices.htm. Talbott notes, "The danger of self-forgetfulness, then, is the danger that we will descend to the level of the computational devices we have engineered—not merely imagining ever new and more sophisticated automatons, but reducing ourselves to automatons."

[37] Sherry Turkle. *Alone Together* (Basic Books, 2011), pp. 20-21.

[38] Sherry Turkle. *Alone Together* (Basic Books, 2011), p. 24.

[39] See http://en.wikipedia.org/wiki/October_sky for a description of the movie and the subsequently renamed book by Homer Hickam from which the movie originated.

[40] See especially Chapter 8, "Frankenstein's Problem," of Langdon Winner's *Autonomous Technology* (MIT Press, 1977).

[41] See http://www.nytimes.com/2011/02/17/science/17jeopardy-watson.html?pagewanted=all&_r=0 for an account of IBM's Watson performing on the television game show "Jeopardy."

[42] We recognize this assertion has been and will continue to be the subject of much debate in many circles—not just among proponents of artificial intelligence. See Joseph Weizenbaum's *Computer Power and Human Reason: From Judgment to Calculation* (1976), pp. 203, 223.

[43] Fred Reed. "Thoughts On The Meritorious Breaking Of Laws." 22 March 2004. http://www.fredoneverything.net/Hide.shtml

[44] See Part Two below for discussion of what can and does happen with ambiguation of roles in aviation.

[45] Joshua Reeves. "If You See Something, Say Something: Lateral Surveillance and the Uses of Responsibility" in *Surveillance and Society*, 2010, 10(3/4), p. 236. Editor's note: Reeves' use of the word "responsibilization" here is tricky. Reeves' essay makes clear that as government pushes for lateral surveillance, the net effect, if anything, leads to less personal internalized morality, at least the kind of morality that rises above rule-hounding and snitching. The reader should be careful not to conclude that increased digitalization of society is somehow improving citizenship as a whole; sadly, evidence we've derived in researching this book points in the opposite direction.

[46] Reeves, p. 237.

[47] Sherry Turkle. *Alone Together*. (Basic Books, 2011), p. 25.

[48] See http://www.nytimes.com/movies/movie/257811/Radio-Bikini/overview for a description of Robert Stone's 1988 documentary about early atomic bomb testing. Late in this Academy Award-nominated film, Stone shows multiple archived film clips of both US and USSR dialogue about eliminating nuclear weapons across the board. Of course, in a few short years after the era described by Stone's film, rather than eliminating those weapons, the two countries engaged in massive races to produce and stockpile as much nuclear weaponry as possible.

[49] Daniel Chandler. "Technological or Media Determinism." 18 September 1985. http://www.aber.ac.uk/media/Documents/tecdet/tdet05.html

[50] See Greg Richter's article on Newsmax.com, "Poll: Congress Not Overreaching on Obama Scandals," from http://www.newsmax.com/Newswidget/Obama-CNN-poll-McConnell-Pfeiffer/2013/05/19/id/505229?promo_code=F6A9-1&utm_source=triblive&utm_medium=nmwidget&utm_campaign=widget phase1. In addition to the well-known IRS targeting stories unfolding in 2013, Richter claims the "Department of Health and Human Services, the Federal Communications Commission, and the Securities and Exchange Commission all have targeted groups." The SEC?

[51] Reeves, p. 238.

[52] Reeves, p. 242.

[53] Jeremy Packer. "Becoming Bombs: Mobilizing Mobility in the War of Terror." *Cultural Studies*, 2006, 20 (4–5), pp. 378–399.

[54] Jeremy Packer. "Becoming Bombs: Mobilizing Mobility in the War of Terror." *Cultural Studies*, 2006, 20 (4–5), p. 273.

[55] Reeves, p. 243.

[56] See "This Teddy Bear Has Sensors That Measure Your Kid's Health" by Kit Eaton, from http://www.fastcompany.com/3014491/fast-feed/this-teddy-bear-has-sensors-that-measure-your-kids-health for a description of a new level of tech-enabled child monitoring: a teddy bear with the ability to track a child's pulse and blood pressure and report that medical information to a mobile phone or mobile device.

[57] See the description of the tragic L-1011 crash at: http://aviation-safety.net/database/record.php?id=19721229-0

[58] See Wikipedia's description of the tragic L-1011 crash at: http://en.wikipedia.org/wiki/Eastern_Air_Lines_Flight_401#cite_note-2

[59] Langdon Winner. "Complexity, Trust, and Terrorism." *Netfuture*, issue 137, October 22, 2002: "Technology and Human Responsibility," a publication of "The Nature Institute," editor Stephen Talbott: http://netfuture.org/2002/Oct2202_137.html

[60] See MailOnline's "Digital Dementia on the Rise as Young People Increasingly Rely Upon Technology Instead of Their Brain," from http://www.dailymail.co.uk/health/article-2347563/Digital-dementia-rise-young-people-increasingly-rely-technology-instead-brain.html

[61] Stephen L. Talbott. "The Fundamental Deceit of Technology." *Netfuture*. 14 December 1995.

[62] Stephen L.Talbott. "The Fundamental Deceit of Technology." *Netfuture*. 14 December 1995.

[63] Lawrence C. Katz and Gary N. Grubb. "Enhancing US Army Aircrew Coordination Training." *ARI Special Report*, 56, May 2003, p. 1. http://www.dtic.mil/dtic/tr/fulltext/u2/a415767.pdf

[64] Neil Postman. *Technopoly: The Surrender of Culture to Technology* (First Vintage Books Edition, 1992), p. 115.

[65] See "Mind-Reading Robots Coming Ever Closer," from *Science Daily*: http://www.sciencedaily.com/releases/2013/11/131114102558.htm

[66] St. Ephraim's prayer: "Oh Lord and Master of my Life, take from me the spirit of sloth, despair, lust of power and idle talk; but give rather a spirit of chastity, humility, patience and love to thy servant. Yea, Oh Lord and King, grant me to see my own transgressions, and not to judge my brother, for blessed art Thou unto ages of ages, amen."

[67] From http://www.pewstates.org/research/reports/one-in-100-85899374411

[68] See http://www.msha.gov/stats/centurystats/coalstats.asp for the US's frightening, albeit steadily improving, coal safety numbers. See http://www.ncbi.nlm.nih.gov/pmc/articles/PMC3056041/ to note that as recently as 2003, China had nearly 5,000 coalmining deaths in a single year and still routinely loses more than 1,000 men per year in coal mines.

[69] See http://johntaylorgatto.com/underground/ for access to John Taylor Gatto's book. As Adam Robinson, co-founder of *The Princeton Review*, noted, it is "A work of breathtaking scholarship and encyclopedic scope." For our purposes here, note that Gatto shreds any notion that states may, in our epoch, somehow grant themselves a new right to surveillance and control based on the introduction of new technologies alone.

[70] Tom Brignall III, "The New Panopticon: The Internet Viewed as a Structure of Social Control" in *Theory and Science, 2002*. Brignall III describes Bentham's Panopticon as an "architectural algorithm"; Bentham depicts his idea using a physical layout for a prison in which all the inmates could be observed all the time from a central point in the building, but he meant his idea to be applied as a means of social control, not only in prisons, but also in schools, cities, and factories. http://theoryandscience.icaap.org/content/vol003.001/brignall.html

[71] See Katie Glueck, "US Wants to Destroy Privacy Worldwide," from: http://www.politico.com/story/2013/06/glenn-greenwald-us-privacy-92400.html#ixzz2VYWMfqp7

[72] See the article at http://www.crimemuseum.org/library/imprisonment/jeremyBentham.html

[73] Jeremy Bentham. *The Panopticon Writings* (Verso, 1995), pp. 29-95.

[74] U.S. Congress, Office of Technology Assessment, Verification Technologies: Cooperative Aerial Surveillance in International Agreements, OTA-ISC-480 (Washington, DC: U.S. Government Printing Office, July 1991), p. 12.

[75] See Noah Barkin's article, "Germans accuse US of Stasi tactics before Obama visit" from Reuters,

http://www.reuters.com/article/2013/06/11/cnews-us-usa-security-germany-idCABRE95A0T820130611

and Daniel Trotta's article, "At U.N., Brazil's Rousseff blasts U.S. spying as breach of law" from Reuters,

http://www.reuters.com/article/2013/09/24/us-un-assembly-brazil-idUSBRE98N0OJ20130924 --

For a slightly contrasting point of view, see "NSA spying allegations: Are US allies really shocked?" by Jonathan Marcus: http://www.bbc.com/news/world-europe-24676392. Indeed, Marcus's question, posed in the title of his article, brings up the semi-obvious and disturbing point that given how actively all states seem to monitor their neighbors, much of the outrage seen in post-Snowden headlines may be feigned. And if feigned, we are perhaps even more at the mercy of the elites of the world to provide us some vestige of privacy and dignity in our personal relationships and communications than we have thus far realized.

[76] Anne Gearan and Philip Rucker. "Clinton: It 'might have been smarter' to use a State Dept. e-mail account." *The Washington Post*. 11 March 2015. http://www.washingtonpost.com/politics/hillary-clinton-to-answer-questions-about-use-of-private-e-mail-server/2015/03/10/4c000d00-c735-11e4-a199-6cb5e63819d2_story.html

[77] Jamais Cascio in "The Rise of the Participatory Panopticon" from http://www.worldchanging.com/archives/002651.html

[78] Takashi Nakamichi. "Fukushima Operator Probes Whether Rat Caused Blackout." *The Wall Street Journal*. 20 March 2013. http://online.wsj.com/article/SB10001424127887324103504578372030233141850.html

[79] Evgeny Morozov talk at TED 2009, video and text accessed from: http://www.ted.com/talks/evgeny_morozov_is_the_internet_what_orwell_feared.html

[80] Langdon Winner. "Complexity, Trust, and Terrorism." *Technology and Human Responsibility*, 137, October 22, 2002. A publication of the *Nature Institute*, editor Stephen L. Talbott: http://netfuture.org/2002/Oct2202_137.html

[81] Harriet Crawford. "The Dark Side of Meditation and Mindfulness." *Daily Mail*, 22 May 2015. http://www.dailymail.co.uk/health/article-3092572/The-dark-meditation-mindfulness-Treatment-trigger-mania-depression-psychosis-new-book-claims.html

[82] Langdon Winner. "Complexity, Trust, and Terrorism." *Technology and Human Responsibility*, 137, October 22, 2002. A publication of the *Nature Institute*, editor Stephen L. Talbott: http://netfuture.org/2002/Oct2202_137.html

[83] Actually, a close reading of Winner's outstanding body of work indicates he's likely in agreement with our belief that we need an expanded view of

reliability.

[84] See http://www.chron.com/news/medical/article/Robot-hot-among-surgeons-but-FDA-taking-fresh-look-4419667.php

[85] See DECLARATION OF WILLIAM E. BINNEY IN SUPPORT OF PLAINTIFFS' MOTION FOR PARTIAL SUMMARY JUDGMENT REJECTING THE GOVERNMENT DEFENDANTS' STATE SECRET DEFENSE, from https://publicintelligence.net/binney-nsa-declaration/

[86] Tim Rayner. "Foucault and Social Media: Life in a Virtual Panopticon" in *Philosophy for Change*, from http://philosophyforchange.wordpress.com/2012/06/21/foucault-and-social-media-life-in-a-virtual-panopticon/

[87] Rayner.

[88] "Sousveillance" itself is a neologism meaning surveillance from the bottom up. In the US, the Rodney King incident provides an early example of technology-based sousveillance. More recently, Julian Assange and Edward Snowden have taken sousveillance to new extremes; time will tell if their actions are simply a part of a Hegelian dialectic that will result in even more broadly "accepted" surveillance. Indeed, early reactions in the US intelligence communities to Snowden's revelations seem more focused on tightening internal security (i.e., increasing internal surveillance and monitoring) than they do on limiting surveillance in general.

[89] Fred Reed. "Thoughts On The Meritorious Breaking Of Laws." 22 March 2004. http://www.fredoneverything.net/Hide.shtml

[90] We note potential contradiction and/or hypocrisy here—the same sort of contradiction the founding fathers of the US political system addressed via separation of government powers: believing, with Hobbes and Bentham, that men cannot be trusted with power normally evokes a search for royal blood, a search for Ubermenschen, or attempts to employ political systems so effective that feeble, greedy men can engage those systems without the need for eviscerating rights and privacy. When Fred Reed (or we) seek human freedom in an American context, hoping to enjoy an expression of that freedom by limiting government surveillance, we simultaneously admit that men and power are a dangerous mixture and request that common citizens like us be entrusted with a variant of that mixture. This apparent contradiction—trusting the untrustable—is at the core of the original US political system and can be seen in many acts of leadership (not least of which, we include acts of parenthood and instruction of the young).

[91] In fairness to Langdon Winner and his many profound insights, he voices concerns that begin, fairly enough, with a desire that the technologies we use act reliably—in the traditional sense of the word "reliable." He goes far beyond this hopefully obvious point about technology to explore our

many weaknesses and our sometimes profound blindness regarding the multitude of situations in which technology may betray us—situations that extend far beyond *task-specific* failures of technology.

[92] For just one of thousands of possible examples of harm caused by a strong ability to surveil, see "NYers furious over photos taken through windows" from *The Associated Press*: http://www.atlanticbb.net/news/read/category/Top%20News/article/ap-nyc_artists_secret_photos_raise_privacy-ap

[93] Morozov, p. 63.

[94] A description of the violent incident with Rodney King and the Los Angeles Police Department—recorded via video camera by a citizen and subsequently replayed around the world—can be found in an online version of King's biography: http://www.biography.com/people/rodney-king-9542141

[95] Tom Brignall III. "Theory and Science" in *The New Panopticon: The Internet Viewed as Structure of Social Control* (2010). http://theoryandscience.icaap.org/content/vol003.001/brignall.html

[96] See "Sony imagines 'Smart Wig' to monitor health, give directions, and read facial expressions," from http://www.theverge.com/2013/11/21/5129554/bizarre-sony-smartwig-patent-turns-wigs-into-wearable-computing-device

[97] Joshua Reeves. "If You See Something, Say Something: Lateral Surveillance and the Uses of Responsibility." *Surveillance and Society*, 2012, 10(3/4), pp. 235-248.

[98] Reeves, p. 238.

[99] See Julian Sanchez's article, "Is Lessig's Free Culture Just a Modern Das Copyright?" in http://arstechnica.com/uncategorized/2008/04/is-lessigs-free-culture-just-a-modern-das-kopyright/

[100] Reeves, p. 241.

[101] Reeves, in "If You See Something, Say Something: Lateral Surveillance and the Uses of Responsibility," notes on p. 239 that David Garland, Glen Burchell, and Nikolas Rose have used the term "responsibilization" to describe government initiatives such as Janet Napolitano's "If You See Something, Say Something" campaign.

[102] Morozov, p. 80.

[103] Morozov, p. 81.

[104] See Julie Rehmeyer, "Do subatomic particles have free will?" from http://www.sciencenews.org/view/generic/id/35391/description/Do_sub atomic_particles_have_free_will

[105] Onora O'Neill. *A Question of Trust* (Cambridge University Press, 2002), pp. 72-73.

[106] Reeves, p. 236.

[107] See the Wikipedia article "Get Smart," from

http://en.wikipedia.org/wiki/Get_smart

[108] See Mike Masnick's 20 Nov 2013 article, "FISA Court Tells The DOJ That It Needs To Explain Why It's Ignoring Order To Declassify Surveillance Opinion," from http://www.techdirt.com/articles/20131120/10544925304/fisa-court-tells-doj-that-it-needs-to-explain-why-its-ignoring-order-to-declassify-surveillance-opinion.shtml

[109] Karl Popper. *The Open Society and Its Enemies* (Princeton University Press, 1994).

[110] Editing notes from Michael Dell, received 6 June 2016

[111] See Sonali Kohli, "Bill Gates Joins Elon Musk and Stephen Hawking in Saying Artificial Intelligence is Scary," *Quartz*, from: http://qz.com/335768/bill-gates-joins-elon-musk-and-stephen-hawking-in-saying-artificial-intelligence-is-scary/

[112] Nicholas Carr. "Why Skepticism is Good: My Reply to Clay Sharkey," from: http://www.britannica.com/blogs/2008/07/why-skepticism-is-good-my-reply-to-clay-shirky/ -- Nicholas Carr is a member of Britannica's Editorial Board of Advisors and the author, most recently, of *The Big Switch: Rewiring the World, from Edison to Google*

[113] Nicholas Carr. "Is Google Making Us Stupid?" *The Atlantic,* July/August 2008. http://www.theatlantic.com/magazine/archive/2008/07/is-google-making-us-stupid/306868/

[114] See FoxNews.com's "Digital dementia: The memory problem plaguing teens and you adults," from http://www.foxnews.com/health/2013/08/15/digital-dementia-memory-problem-plaguing-teens-and-young-adults/, and Dr. Sylvia Hart Fredj's "The Rise of Digital Dementia: Protecting Yourself and Your Clients," from the American Association of Christian Counselors, accessed at http://www.aacc.net/2013/10/18/the-rise-of-digital-dementia-protecting-yourself-and-your-clients/, and Elijah Wolfson's "Digital Dementia on the Rise in South Korea; Childhood Internet Addiction Must Be Addressed, Experts Say," in the *Medical Daily*, accessed at http://www.medicaldaily.com/digital-dementia-rise-south-korea-childhood-internet-addiction-must-be-addressed-experts-say-247100

[115] See Clay Shirky's "Why Abundance is Good: A Reply to Nick Carr," in *Encyclopedia Britannica Blog* at: http://www.britannica.com/blogs/2008/07/why-abundance-is-good-a-reply-to-nick-carr/

[116] See "Hegelianism" in *New World Encyclopedia*: http://www.newworldencyclopedia.org/entry/Hegelianism. Also, as noted in our key definition of Hegelianism, at least one of our reviewers believes we may not have captured the essence of the teachings of Hegel here. While

we accept that criticism, we also note that if folks apply theories in certain ways, and they do so consistently, it may be irrelevant that they originally misunderstood any given theory. Sadly, the manipulation of those who embrace Hegel's teaching, in whatever perverted form it may have been misunderstood, is well-documented in the wars and conflict history of the 20th century.

[117] See Nick Bilton, "Disruptions: New Motto for Silicon Valley: First Security, Then Innovation," published in the *New York Times* on 5 May 2013: http://bits.blogs.nytimes.com/2013/05/05/disruptions-new-motto-for-silicon-valley-first-security-then-innovation/?WT.mc_id=NYT-E-YHO-NYT-E-BIZ-050613-L1&nl=el

[118] For one of many examples of this behavior on the part of the US government, see "President-elect Urges Electronic Medical Records in Five Years," by ABC News: http://abcnews.go.com/Health/President44/story?id=6606536&page=1#.Uc9K2euvXs0

[119] Without naming specific companies, more than one software company has sold billions upon billions of dollars' worth of "fixes" to their existing flawed products, sometimes even building the expectancy of such a sell/break/repair process into their marketing plans, e.g., software leasing.

[120] For an outspoken and bluntly worded discussion of Hegelianism in our lives, see "The Devil's Logic," by John J. Parsons: http://www.hebrew4christians.com/Articles/Hegelianism/hegelianism.html

[121] Twitter enabled two-factor authentication in the spring of 2013. See https://www.eff.org/deeplinks/2013/05/howto-two-factor-authentication-twitter-and-around-web, accessed 26 August 2016.

[122] See the *Economist* article, "Planned Obsolescence" for a discussion of its intentional use in multiple industries, including of course the software industry: http://www.economist.com/node/13354332. The article was adapted from *The Economist Guide to Management Ideas and Gurus* by Tim Hindle (Profile Books, 2008).

[123] Jeffrey Brady. "Amish Community Not Anti-technology, Just More Thoughtful." *National Public Radio*, aired 2 Sep 2013. Accessed online at: http://www.npr.org/templates/transcript/transcript.php?storyId=217287028

[124] Richard M. Weaver. *Ideas Have Consequences*. From the Introduction, accessed at: http://www.orthodoxytoday.org/articles5/WeaverIdeas.php

[125] Breaks from techno-centrism sometimes show up in the most unlikely places and don't necessarily involve days' or weeks' worth of absence from a job. Google—the company—apparently has, among its many unusual and interesting diversions for employees—a Japanese meditation spot accompanied by signage saying, "Experience the joy and mindfulness of just

eating, try monotasking for lunch today." For those unfamiliar with the caffeine-injected culture of Information Technology, this meditation spot is an invitation to set down the cell phone, iPad, laptop, pager, etc, etc, ad infinitum, and simply eat, breath, and think. What a concept—although in fairness we should note that eating in this fashion has been done before. From an article in *The Independent* by Ian Burrell, accessed from: http://www.independent.co.uk/life-style/gadgets-and-tech/features/inside-google-hq-what-does-the-future-hold-for-the-company-whose-visionary-plans-include-implanting-a-chip-in-our-brains-8714487.html

[126] Richard M. Weaver. *Ideas Have Consequences*. From the Introduction, accessed at: http://www.orthodoxytoday.org/articles5/WeaverIdeas.php

[127] Again, since our writing here is not an extended Luddite rant, but a book written with some of the most current computer-assisted word processing available, we deem Weaver's comments relevant and even foundational for some of our arguments about digital reliability in general.

[128] Jacques Ellul. *BrainyQuote.com*: http://www.brainyquote.com/quotes/authors/j/jacques_ellul.html

[129] New One-volume Edition, pp. xlii and xliii of the Introduction. Princeton University Press, 1994.

[130] The authors of Harvard University's *Don't Panic—Making Progress on the 'Going Dark' Debate* note, "A service, which entails an ongoing relationship between vendor and user, lends itself much more to monitoring and control than a product, where a technology is purchased once and then used without further vendor interaction." From https://cyber.law.harvard.edu/pubrelease/dont-panic/Dont_Panic_Making_Progress_on_Going_Dark_Debate.pdf. Regarding government/corporate use of incrementalism, see "Surprise! NSA Data Will Soon Routinely Be Used for Domestic Policing that Has Nothing to Do with Terrorism," from https://www.washingtonpost.com/news/the-watch/wp/2016/03/10/surprise-nsa-data-will-soon-routinely-be-used-for-domestic-policing-that-has-nothing-to-do-with-terrorism/

[131] The Modern Library Board selected Popper's book as one of the 100 Best Nonfiction books of the 20th century in 1999.

[132] Karl Popper. *The Open Society and Its Enemies—The Spell of Plato,* vol. I (Butler and Tanner, 1945), p. 574.

[133] Popper, p. 56. "But there are other important sociological laws, connected with the functioning of social institutions. These laws play a role in our social life corresponding to the role played in mechanical engineering by, say, the principle of the lever. For institutions, like levers, are needed if we want to achieve anything which goes beyond the power of our muscles. Like machines, institutions multiply our power for good and evil."

[1134] Alexis de Tocqueville. *Democracy in America*, Book Four, Chapter VI, accessed in digital format from:
http://xroads.virginia.edu/~HYPER/DETOC/ch4_06.htm

[1135] Mike Weston. "'Smart Cities' Will Know Everything About You." *The Wall Street Journal.* http://www.wsj.com/articles/smart-cities-will-know-everything-about-you-1436740596

[1136] Langdon Winner. "Are Humans Obsolete?"
http://homepages.rpi.edu/~winner/AreHumansObsolete.html

[1137] Oswald Spengler. *Man and Techniques: A Contribution to a Philosophy of Life*, 2nd edition (Arktos Media, 2015), p. 25.

[1138] De Tocqueville. *Democracy in America*, p. 1 of the "Author's Preface," digital version published at
http://xroads.virginia.edu/~HYPER/DETOC/preface.htm

[1139] See "ENIAC" at Wikipedia.org, from
https://en.wikipedia.org/wiki/ENIAC. An interesting side note about the reliability of the world's "first electronic general purpose computer": "Several tubes burned out almost every day, leaving it nonfunctional about half the time. Special high-reliability tubes were not available until 1948. Most of these failures, however, occurred during the warm-up and cool-down periods, when the tube heaters and cathodes were under the most thermal stress. Engineers reduced ENIAC's tube failures to the more acceptable rate of one tube every two days. According to a 1989 interview with Eckert, 'We had a tube fail about every two days and we could locate the problem within 15 minutes.' In 1954, the longest continuous period of operation without a failure was 116 hours—close to five days."

[1140] Oswald Spengler. *Man and Technics*, p. 27.

[1141] Oswald Spengler. *Man and Technics*, p. 28.

[1142] Oswald Spengler. *Man and Technics*, p. 29

[1143] Oswald Spengler. *Man and Technics*, p. 37.

[1144] Oswald Spengler. *Man and Technics*, p. 63.

[1145] Oswald Spengler. *Man and Technics*, pp. 66-67.

[1146] Oswald Spengler. *Man and Technics*, p. 77.

[1147] Oswald Spengler. *Man and Technics*, p. 77.

[1148] Readers may note that we're mixing determinism and reification in some of the arguments above. This is because in practice, it seems that those who reify history, or nature, the market, technology, etc., with rare exception, assert that the future, as directed by these reified "forces," is pre-determined.

[1149] Richard Stallman. "Cloud Computing Is a Trap, Warns GNU Founder Richard Stallman." *The Guardian*:
http://www.guardian.co.uk/technology/2008/sep/29/cloud.computing.richard.stallman

[1150] Steve Watson. "Transhumanist Kurzweil Predicts Human/Computer

Hybrids by 2030s." *PrisonPlanet.com*:
http://www.prisonplanet.com/transhumanist-kurzweil-predicts-
humancomputer-hybrids-by-2030s.html

[150] David Hanson. "America in 2025: Bill Gates' Scary Prediction." *The Motley Fool*: http://www.fool.com/ecap/stock-advisor/gates-
prediction/?utm_source=yahoo&utm_medium=contentmarketing&utm_c
ampaign=advertorial_lg&paid=8484&psource=esayho7410860072&waid=
8482&wsource=esayhowdg0860228

[152] Such behavior in a religious milieu would be granted "cult" status at best.

[153] From: http://trueworldhistory.info/docs/quotes.html

[154] Postman. *Technopoly*, p. 130.

[155] Martin Shapiro. "Judges as Liars." *Berkeley Law Scholarship Repository*:
http://scholarship.law.berkeley.edu/cgi/viewcontent.cgi?article=1265&con
text=facpubs, accessed 2 April 2016. Shapiro references Thomas W.
Merrill, A Modest Proposal for a Political Court, 17 Harv. J.L. & PUB.
POL'Y 137 (1994).

[156] For a definition of the phrase "immanentization of the eschaton," see
the thoughts and writings of Eric Voegelin and William F. Buckley, Jr.

[157] See Norbert Wiener, as quoted by Neil Postman, *Technopoly: The Surrender of Culture to Technology*, First Vintage Books Edition, p. 115.

[158] "NSA Official: 'We are Now in a Police State,'" from
http://www.cnsnews.com/mrctv-blog/matt-vespa/nsa-official-we-are-
now-police-state

ABOUT THE AUTHORS

Jeff Krinock

Jeff is a former USAF helicopter and fighter pilot. His background is diverse, having studied (and, far too rarely, mastered) subjects ranging from theology to aviation to enology, human relations, creative writing, and back again. His post-USAF career has included stints as either a consultant to or employee of organizations such as Hewlett-Packard, Microsoft, the Army Medical Command (AMEDD), Joint Forces Command (JFCOM) and IBM. His first love is writing, which he views as his chosen vocation, but which he is willing to woo and wed just as readily in the form of an avocation.

Matt Hoff

Matt Hoff is a consultant with a global technology company and he has worked with a variety of both private and government clients. Matt studied Computer Science at Michigan Technological University and has a Masters in Educating Adults from DePaul University. Matt's interest in technology started at a young age, when his father brought home a Tandy 1000 SX – his family's first home computer.

www.ingramcontent.com/pod-product-compliance
Lightning Source LLC
Chambersburg PA
CBHW071128050326
40690CB00008B/1377